NATURAL ACTS

ALSO BY DAVID QUAMMEN

NATURAL ACTS

A SIDELONG VIEW OF SCIENCE & NATURE

Revised and expanded edition, with a new introduction
Including "Planet of Weeds" and the Megatransect series

David Quammen

W. W. NORTON & COMPANY
New York   London

To M.E.Q. and W.A.Q.
for everything

Manufacturing by The Courier Companies, Inc.
Book design by Wesley Gott
Production manager: Andrew Marasia

Library of Congress Cataloging-in-Publication Data

Quammen, David, 1948–
Natural acts : a sidelong view of
science & nature / David Quammen. —
Rev. and expanded ed., with a new introduction including
"Planet of weeds" and the Megatransect series.
p. cm.
Includes bibliographical references and index.
ISBN 978-0-393-05805-5 (hardcover)
1. Natural history—Miscellanea. I. Title.
QH45.5.Q36 2008
508—dc22

2007027545

W. W. Norton & Company, Inc., 500 Fifth Avenue, New York, N.Y. 10110
www.wwnorton.com

W. W. Norton & Company Ltd.Castle House, 75/76 Wells Street, London W1T 3QT

1 2 3 4 5 6 7 8 9 0

CONTENTS

Learning Curve

THIS BOOK AS YOU HOLD IT is a chimerical creature, like a griffin, bird-shaped in front with a mammalian caboose. It consists of two asymmetrical but, I hope, complementary halves: a selection of what I take to be the most durable of my recent shorter nonfiction (in the fourth section, titled "After Thoughts") and a selection of my earliest work in roughly the same vein (sections one through three), most of which appeared in the first edition of *Natural Acts*, published in 1985. Combining them now in one volume is probably risky, and perhaps presumptuous, but I'd like to think it serves three modest purposes: 1) reviving the best parts of a book that is otherwise out of print, 2) putting back into circulation some recent essays on subjects about which I have strong convictions (such as "Planet of Weeds"), and 3) offering readers evidence by juxtaposition of how one writer might have changed and developed over a period of twenty-six years.

Once I was a young man so blithe and unfettered that I could write the sentence "Biology has great potential as vulgar entertainment." (See the following "Old, Ingenuous Introduction.") It was like being a sleek juvenile shark before the remoras of sophistication and judiciousness attached themselves. I had, in those years, only recently blundered into the craft of science journalism. I was unencumbered by experience, professional qualifications, broad knowledge, or a sense of decorum. I had no training in science, but then again I had no training in journalism either.

It all began in the winter of 1980–81, when I wrote a short essay for *Outside* magazine on the redeeming merits, insofar as there are any, of mosquitoes. I had pitched the idea to the magazine's editor, John Rasmus, after a long day of fly-fishing in the small Montana town where I then lived. Mr. Rasmus, an august figure (it seemed to me) but even younger than I, was visiting Montana to find and cultivate new voices. Along with Rasmus and a close friend (Stephen Byers, nowadays a New York editor himself), I took off in a johnboat to torment trout on our local stretch of river. After the big rainbows had been subdued and released and the sun had set behind the Gravelly Mountains, Steve and his wife of the time (E. Jean Carroll, now a columnist for *Elle*) and I softened John up with a ranch-kitchen dinner of whiskey and steak and whiskey. Then I made my pitch: What about a piece on mosquitoes? The upside! The counterintuitive, good things to be said for those noxious insects! Um, okay, said poor Rasmus. Although the whiskey wore off within a day or two, the deal stuck.

My mosquito paean was intended to be a one-off piece. But sometime that winter, after receiving a draft, John called me with an unexpected proposition: Would I be interested in becoming a columnist for *Outside*, writing regularly on nature and science under the column title (already deployed in the magazine by a previous writer) "Natural Acts"? The mosquito piece, he suggested, could run as the first of my columns and I could follow in the same vein, with whatever nature-related expository and opinionated jive I cared to offer. At that moment, just beginning my efforts at free-lancing, I could scarcely go to the grocery store and buy hamburger without first balancing my checkbook to see exactly where I stood. Yes, I said. Yes, absolutely, I'll be glad to do it . . . for a year or two.

Fifteen years passed and, son of a gun, I found I had written about 160 columns. During that time I had been given extraordinary freedom and trust by John Rasmus and his successors, indulged to follow my curiosity virtually wherever it led (so long as each monthly essay had some connection to nature or science) and to educate myself somewhat in the fields of ecology, field biology, and evolutionary theory. I was still an outsider to the biological sciences, a nonexpert with a noneducation, but those areas had become familiar to me as a journalistic beat. You don't have to be a cop or a burglar to cover the crime stories down at the courthouse, and you don't have to be a biologist to write about biology. My lack of formal scientific training may even have been an advantage in some ways, leaving me with a fresh eye and an ingenuous ignorance similar to those of the general readers as whose proxy I tried to serve.

My style of research for the early columns, partly conditioned by the fact that I'd never darkened the door of a school of journalism, was roundabout, unsystematic, ruminative as a feral goat, highly dependent on arcane printed materials and personal experience, and, some might say, unprofessional. That is, I didn't choose a subject, then choose an angle, then call experts on the phone and ask for quotes. Instead, I wandered amid obscure sources (old books, back issues of scientific journals, newspaper clippings, classic texts, fragments of remembered literature) and, whenever possible, wild places (the Amazon headwaters, the Sonoran Desert, the rivers of western Montana) until my attention seized on a fact or a creature that seemed to me both interesting and important. I researched outward from that point, trying to find a broader cultural context, and also downward, more deeply into it; then, usually under severe deadline pressure, I wrote a short essay in a tone of voice meant to seem as immediate and informal as a telephone harangue to my best friend. By embracing the results, the readers of *Outside* during those years subsidized and encouraged my overdue scientific education. I take this chance to thank them again.

The book *Natural Acts*, as originally published in 1985, collected about two dozen of those early columns, along with a few longer pieces I had written for *Esquire, Audubon,* and other magazines. The good people at Nick Lyons Books (and that company's successor, Lyons and Burford) kept it available in hardback over the years, while a couple of paperback editions came and went. Meanwhile, as I continued writing columns, the character of those short essays gradually changed, becoming (for better or worse) more careful of fact and complexity, more deeply informed by the main themes and principles of ecology and evolutionary biology, possibly more shapely, definitely less naive and blunt. Some of them are gathered in my later collections, such as *The Boilerplate Rhino*. In 1995, I resigned from the columnist's role when I reached such a point of exhaustion—exhaustion not with science or with nonfiction, just with that jaunty little form, the column-length essay—that I felt I might start repeating or imitating myself. I hadn't seen the end coming. I just realized one day, suddenly, as I rewrote a troublesome column, that this was it.

Since then I've divided my working time between nonfiction books and magazine pieces of a sort that wasn't possible within the constraints of a monthly column. Most of the latter have been written for *National Geographic* and *Harper's*, two very different magazines, with very different readerships, that have offered me very different opportunities and satisfactions. One blessing they have both provided is the freedom to spend much more time gathering raw material from the field—out amid the equatorial forests, the snowy mountainsides, and the laboratories where some biologists (see "Clone Your Troubles Away") do their work. Related to that is another enlarged opportunity: to follow a subject much further and deeper, either in terms of sheer miles walked ("The Megatransect") or in terms of complexity, than a column allows. In an extended essay or feature-length story, a writer isn't forced so sternly toward neatness, toward the elliptical comment as opposed to the unfolded thought, toward lapidary concision and punch lines. He has space to walk around a subject, not just up to it.

In the years since I quit the column, I have never regretted ending that hectic and privileged duty. Then again, though, there is something special about a columnist's ongoing relationship with his audience. Almost necessarily, it becomes personal, at least in its one-sided way. Readers turn to a certain page of the magazine month after month, by habit and affinity, expecting not just information but a familiar human voice. They begin to feel that they know you. They want to hear what you've got to say about whatever subject may be in question. This invites an intimacy of tone and an occasional unguarded revelation, a showing of authorial ankle, that can be put to some good uses beyond the legitimate (but perilous) use of sheer autobiography. I accepted that invitation often enough when I was a columnist, maybe too often, and

I've intruded myself as a major character in a few of the longer pieces here also. Never mind which; you'll know me when you see me. Whatever those pieces may try to say or whatever they may inadvertently reveal, I hope they'll suggest that I've learned at least a little something over the years besides science.

One truth I've learned—big surprise—is that magazine writing comes and goes. Most magazines (even those with august histories, such as *Harper's* and *National Geographic*) have an issue-by-issue hold on our attention that's shorter than the refrigerator life of an opened jar of mayonnaise. But where else can short nonfiction—the essay, the profile, the adventure narrative, the thoughtful investigation—find its first existence but in magazines? That's why magazine writers, at least the more presumptuous of us, harbor an unspoken division of purpose: We're writing for the subscribers, yes, and the people in the barbershops and the dental waiting rooms, yes, but we're also writing for time, posterity, the shelf. We're writing for you, who will find us later between stiff covers, without the glossy ads and the resubscription cards.

And that's why I'm gratified to introduce, for your kindly attention, a book of short nonfiction pieces that in large part I introduced once before, twenty-two years ago. I haven't entirely expunged the youthful bumptiousness, but this edition is older and at least a little wiser.

An Inordinate Fondness for Beetles

BIOLOGY HAS GREAT POTENTIAL as vulgar entertainment. For that matter, so do geology, ecology, paleontology, and the history of astronomy. Browsing at the intricacies of the natural world and at the lives and works of the scientists who map that world can be fascinating, mesmeric, outrageous good fun. Alas, it can also be heart-squashingly boring. Edifying but deadly, like the novels of Henry James. Or just harmlessly quaint—nature as vast curio shop with an inventory tending to cuteness. The crucial difference, at least to my biased view, is in the angle of approach. The choice of perspective. Lively writing about science and nature depends less on the offering of good answers, I think, than on the offering of good questions.

My own taste runs toward such as *What are the redeeming merits, if any, of the mosquito?* Or *Why is the act of sex invariably fatal for some species of salmon?* Or *Are crows too intelligent for their station in life? Why do certain bamboo species wait 120 years before bursting into bloom? How do seals stay cool in the Arctic? Does a termite colony constitute many little animals or one big one?* Or, perhaps best of all, *Why are there so many different species of beetle?*

That last question has enthralled me for years—ever since I came across the delightful and mind-opening fact that one of every four animals on Earth (by count of the number of species) is, yes, a beetle.

This is actually a conservative estimate, reflecting only the number of animal species that have been discovered and identified by science. Add up all the known species of mammals and birds and reptiles and amphibians, all the fishes and crustaceans and protoplasmic tentacle-waving sea creatures, every brand of zooplankton that has ever been given a name, every type of worm, every flea every mite every spider, also of course every insect on the current entomological roster, and the total comes to around 1.25 million known species of animal. Of that vast assemblage, one in four is a beetle.

We're talking about an order called Coleoptera, containing 300,000 officially described species. And new beetles are being discovered almost every time some scientist waves a net through a rainforest. (A Smithsonian entomologist named Terry Erwin believes, from his study of jungle canopy in Peru, that there might be as many as 12 *million* species of beetle.)

Each of those 300,000 described species conforms to the basic coleopteran pattern: an insect showing complete metamorphosis (progressing from egg to larva, then pupa, then adult) which in its adult form has biting mouthparts, a pair of front wings drastically modified into hard protective covers (called elytra), a pair of lighter rear wings underneath those covers, and an extraordinarily strong cuticle over the whole body that looks like, and functions as, a suit of armor plating.

Within that basic pattern there is an unimaginable variety of shapes and colors and life strategies—vicious pinchers and rhinoceros horns on the head, anteater snouts, antennae like the most elaborate TV aerial, snapping hinges between thorax and abdomen that allow certain species to turn somersaults, light fixtures for signaling mates after dark, beetles as small as a sesame seed, beetles as large as a mouse, long scrawny beetles and husky broad-shouldered ones, leaf-eaters and fungus-eaters and meat-eaters, some that live underwater in rivers, some that burrow subway tunnels along the cambium layer of trees, some that gather and roll huge Sisyphean balls of dung. They are a very old as well as a very successful group of animals, dating back almost 250 million years, and in that stretch of eons they have had ample time to diversify.

But cockroaches are equally old. So are dragonflies. So are sharks. So are lizards. Why, then, are there so cotton-pickin' many species of Coleoptera? Why 300,000 variations?

I don't know. I don't know of anyone who knows. I'm still waiting for some evolutionary biologist to propose a convincing explanation—but my secret hope is that no one can or will.

Meanwhile the famed British geneticist J.B.S. Haldane has left us a valuable comment on this subject. Besides being an eminent scientist from a family of eminent scientists, Haldane was well known in the 1930s as a Marxist and a curmudgeon. Oral tradition among biologists records that Haldane was once cornered by a group of theologians. One of them asked what inferences a person could draw from a study of the created world as to the nature of the Creator. Haldane answered: "An inordinate fondness for beetles."

As it happens, J.B.S. Haldane also believed that science has great potential as vulgar entertainment. During the 1930s and 1940s he wrote a long series of science essays for the general public, most of which appeared in the *Daily Worker*. These short pieces had titles such as "Why I Admire Frogs," "Living in One's Skeleton," "Some Queer Beasts," and (following a case of theft from the British Museum, for which one entomologist went to prison) "Why Steal Beetles?" In spending his considerable wit and his precious working time to produce hundreds of popular essays, Haldane was perhaps the first in a tradition that is now burgeoning: the tradition of scientists who write graceful and accessible essays on scientific subjects for a lay readership.

Loren Eiseley continued that tradition, and today it is rich with the work of Stephen Jay Gould, Lewis Thomas, Freeman Dyson, Alan Lightman, Robert S. Desowitz, and others. I would love to be able to claim a modest toehold in the same tradition. But I can't and I don't. Because I'm not a scientist.

What I am is a dilettante and a haunter of libraries and a snoop, the sort of person who has his nose in the way constantly during other people's field trips, asking too many foolish questions and occasionally scribbling notes. My own formal scientific training has been minuscule (and confined largely to the ecology of rivers).

Gould and Thomas and Lightman actually *do* science, in addition to writing about it. I merely *follow* science. In my other set of pajamas I'm not a biologist but a novelist.

This autobiographical information is offered not because I imagine it has any inherent interest but in a spirit of disclaimer, an effort at truth in packaging. The following is not a diet book nor a detective novel nor a collection of essays by a reputable scientist. Nor is it, for that matter, a string of straightforward dispatches from a "science reporter." It is the work of an outsider who is broadly curious but who can never remember the difference between meiosis and mitosis, who has nevertheless been invited to write on scientific subjects by a small number of charming but gullible magazine editors, who tries hard to keep the facts straight, who is not shy about offering opinions, and whose purpose in these pieces has been divided about equally between edification and vaudeville.

In the course of pondering what to say in this introduction, I invented an old saying that goes: "Put a magazine writer between hard covers, and immediately he thinks he's an essayist."

Of course I'm no exception. In defense of that claim, I can say only that 1) the first section of this book, "All God's Vermin," was taking form in my head as a sequence of essays long before I began making my living from magazine work, and 2) I have tried to shape nearly all of these pieces as essays more than as features or profiles or articles, because old magazines go to the recycling center, whereas old books of essays are allowed to turn yellow with dignity on the shelf. (This doesn't apply to the piece called "The Excavation of Jack Horner," which is clearly a profile, commissioned by *Esquire* as such.) But please don't ask me to define "essay," because the atmosphere could quickly grow ponderous. Um, it's a filigreed editorial whose author doesn't know just which side he has argued until he reads the rough typescript. It's a small wobbly verbal dirt bike used for exploring the intellectual backcountry, modest in horsepower yet under imperfect control of the cyclist. See what I mean about the atmosphere?

During the same time span from which these pieces come, I also wrote others that were definitely not essays—they were profiles or articles or reviews. Them you are being spared. The recycling center has swallowed them already.

More important than categorizing the pieces of this book, though, is saying what unites them: subject matter and point of view. The subject matter is nature and the nature of science, with excursions into freshwater biology, geology, entomology, theoretical ecology, the history of astronomy, preservation issues, the role of bats in literature. The point of view is generally oblique and (or so I flatter myself) counterintuitive. My ambition has been to offer some small moments of constructive disorientation in the way nature is seen and thought about. Along the way I have been drawn in particular toward certain creatures that are conventionally judged repulsive, certain places that are conventionally judged desolate, certain humans and ideas that are conventionally judged crazy.

I have also found more fascination in the questions than in the answers. And here's a further question. On what grounds might we assume or hope that—despite the awesome puissance of modern science—any fertile mysteries still abide, unsolved, in the natural world? I can think of 300,000 reasons.

ALL GOD'S
VERMIN

Sympathy for the Devil

UNDENIABLY THEY HAVE A LOT TO ANSWER FOR: malaria, yellow fever, dengue, encephalitis, and the ominous tiny whine that begins homing around your ear just after you've gotten comfortable in the sleeping bag. All these griefs and others are the handiwork of that perfidious family of biting flies known as the Culicidae—the mosquitoes. They assist in the murder of millions of humans each year, carry ghastly illness to millions more, and drive not a few of the rest of us temporarily insane. They are out for blood.

Mosquitoes have been around for 50 million years, which has given them time to figure all the angles. Judged either by sheer numbers or by the scope of their worldwide distribution or by their resistance to enemies and natural catastrophe, they are one of the great success stories on the planet. They come in 2,700 different species. They inhabit almost every land surface, from Arctic tundra to downtown London to equatorial Brazil, from the Sahara to the Himalayas, though best of all they like tropical rainforests, where three quarters of their species reside. Mosquitoes and rainforests, in fact, go together like gigolos and bridge tournaments, pickpockets and camel markets, insurance salesmen and . . . But wait, I was talking about insects.

They hatch and grow to maturity in water, any entrapment of quiet water, however transient or funky. A soggy latrine, for instance, suits them fine. The still edge of a crystalline stream is fine. In the flooded footprint of an elephant, you might find a hundred larval mosquitoes. At that stage of life, as inoffensive juveniles, they use facial bristles resembling cranberry rakes to comb such waters for planktonic food; but on attaining adulthood, they are out for blood.

Blood: It isn't a necessity for individual survival, just a dietary prerequisite of motherhood. Male mosquitoes do not even bite. A guy mosquito lives his short, gentle adult life content, like a swallowtail butterfly, to sip nectar from flowers. As with black widow spiders and mantids, it is only the female that is fearsome. Make of that what larger lessons you dare.

She relies on the blood of vertebrates—mainly warm-blooded ones but also sometimes reptiles and frogs—to finance, metabolically, the development of her eggs.

A female mosquito in a full lifetime will lay about ten separate batches of eggs, roughly 200 in a batch. That's a large order of ovular mass to be manufactured in one wispy body, and to manage it the female needs a rich source of protein; the sugary juice of flowers will deliver quick energy to wing muscles, but it won't help her build 2,000 new bodies. So she has evolved a hypodermic proboscis and learned how to steal protein in one of its richest forms, hemoglobin. Among some mosquito species, the female's first brood will develop before she has tasted blood, but after that she too must have a bellyful for each set of eggs coming to term.

When she drinks, she drinks deeply: The average blood meal amounts to two and a half times the original weight of the insect. Picture Audrey Hepburn sitting down to a steak dinner, getting up from the table weighing 350 pounds, and then flying away. In the Canadian Arctic, where species of the genus *Aedes* emerge in savage, sky-darkening swarms like nothing seen even in the Amazon and work under pressure of time because of the short summer season, an unprotected human could be bitten 9,000 times per minute. At that rate, a large man would lose half his total blood in two hours. Arctic hares and reindeer move to higher ground or die. And sometimes solid mats of *Aedes* will continue sucking the cool blood from a carcass.

Evidently the female tracks her way to a blood donor by flying upwind toward a source of warmer air, or toward air that is both warm and moist or that contains an excess of carbon dioxide, or a combination of all three. The experts aren't sure. Perspiration, involving both higher skin temperature and released moisture, is one good way to attract her attention. In certain villages of Italy there was a folkish belief that to sleep with a pig in the bedroom was to protect oneself from malaria, presumably because the pig, operating at a higher body temperature, would be preferred by mosquitoes. And at the turn of the twentieth century, Professor Giovanni Grassi, then Italy's foremost zoologist, pointed out that garrulous people seemed to be bitten more than those who kept their mouths shut. The experts aren't sure, but the Italians are full of ideas.

Guided by CO_2 or idle chatter or distaste for pork or whatever, a female mosquito lands on the earlobe of a human, drives her proboscis (actually a thin bundle of tools that includes two tubular stylets for carrying fluid and four serrated ones for cutting) through the skin, and gropes with it until she taps a capillary, and then an elaborate interaction begins. Her saliva flows down one tube into the wound, retarding coagulation of the spilled blood and provoking an allergic reaction that will later be symptomized by itching. A suction pump in her head draws blood up the other tube, a valve closes, another pump pulls the blood back into her gut. And that alternate pumping and valving continues quickly for three orgiastic minutes, until her abdomen is stretched full like a great bloody balloon or a fast human hand ends her maternal career, whichever comes first.

But in the meantime, if she is an individual of the species *Anopheles gambiae* in Gabon, the protozoa that cause malaria may be streaming into the wound with her saliva, heading immediately off to set up bivouac in the human's liver. Or if she is *Aedes aegypti* in Peru, she may be drooling out an advance phalanx of the yellow fever virus. If she is *Culex pipiens* in Malaysia, long tiny larvae of filarial worms may be squirting from her snout like a stage magician's spring-work snakes, dispersing to breed in the unfortunate person's lymph nodes and eventually clog them, causing elephantiasis.

No wonder, then, that in the inverted rogue's pantheon of those select creatures not only noxious in their essential character but furthermore lacking any imaginable forgiving graces, the Culicidae are generally ranked below even the deer tick, the lake leech, the botfly, the wolverine, and the black toy poodle. The mosquito, says common bias—and on this the experts tend to agree—is an unmitigated and irredeemable pest.

But I don't see it that way. To begin with, the Culicidae family is not monolithic, and it does have—even from the human perspective—its beneficent representatives. In northern Canada, for instance, *Aedes nigripes* is an important pollinator of arctic orchids. In Ethiopia, *Toxorhynchites brevipalpis* as a larva preys voraciously on the larvae of other mosquitoes, malaria carriers, and then metamorphoses into a lovely, huge, iridescent adult that, male or female, drinks only plant juices and would not dream of biting a human.

But even discounting these innocent aberrations, and judging it only by its most notorious infamies, the mosquito is taking a bad rap. It has been victimized, I submit to you, by a strong case of anthropocentric bias. In fact, the little sucker can be viewed, with only a small bit of squinting, as one of the great ecological heroes of planet Earth. If you consider rainforest preservation.

The chief point of blame, with mosquitoes, happens also to be the chief point of merit: They make tropical rainforests, for humans, virtually uninhabitable.

Tropical rainforest constitutes by far the world's richest and most complex category of terrestrial ecosystem, a boggling entanglement of life-forms and habits and physical conditions and relationships. Those equatorial forests—mainly confined to the Amazon, the Congo basin and its neighboring Central African drainages, the wetter and warmer parts of Indonesia and northern Australia, and parts of mainland Southeast Asia—account for only a small fraction of Earth's surface but serve as home for an inordinate share of our planet's total plant and animal species, including about 2,000 kinds of mosquito. But rainforests lately—in case you've been stuck in an elevator for twenty years and haven't heard—are under siege.

They are being clear-cut for cattle ranching, nibbled away for subsistence agriculture, mowed down with bulldozers and pulped for paper, hacked and dried for firewood, milled into chopsticks and cheap plywood, gobbled up hourly for the sake of "development" in all its ambivalent forms. The current rate of loss, by one rough estimate, amounts to eight acres of rainforest gone, *poof,* since you began reading this sentence. Within a few generations, at that pace, the Amazon will look like New Jersey. Conservation groups are raising a clamor, tossing money at the problem, and making efforts to offer mitigating alternatives, while some governments in the countries at issue take steps for marginal preservation in the form of reserves or national parks. But no one and no thing has done more to delay this catastrophe, over the past 10,000 years, than the mosquito.

The great episode of ecological disequilibrium that we call human history began, so the fossils tell us, in equatorial Africa. Then quickly the focus of intensity shifted elsewhere. What deterred mankind, at least to a large degree, for a very long time, from hacking space for our farms and cities out of the tropical forests? Yellow fever did, and malaria, dengue, filariasis, o-nyong-nyong fever.

Clear the vegetation from the brink of a jungle waterhole, move in with tents and cattle and Jeeps, and *Anopheles gambiae*, not normally native there, will arrive within a month, bringing malaria. Cut the tall timber from five acres of rainforest, and species of viral-transmitting *Aedes*—which would otherwise live out their lives in the high forest canopy, passing yellow fever between monkeys—will fall on you and begin biting before your chainsaw has cooled. Nurturing not only more species of snake and bird than anywhere else on Earth, but also more forms of disease-causing microbe and more mosquitoes to carry them, tropical forests are elaborately booby-trapped against disruption.

The resident forest peoples, living at low densities, gradually acquired some immunity to these diseases, and their hunting-and-gathering economies, grounded in relatively simple technology, minimized their exposure to mosquitoes that favored the canopy or disturbed landscape. Meanwhile the occasional white interlopers, the agents of empire, remained vulnerable. West Africa in high colonial days became known as "the white man's grave."

So as Europe was being stripped of its virgin woods, and India and China, and the North American heartland, the tropical rainforests largely escaped, lasting into the late twentieth century—with some chance, at least, that they may endure a bit longer. Thanks to what? To a concatenation of accidental and deterministic factors, no doubt, among which should be included this: 10 million generations of jungle-loving, disease-bearing, bloodsucking mosquitoes—the Culicidae, nature's Vietcong.

Has Success Spoiled the Crow?

ANY PERSON WITH NO STEADY JOB and no children naturally finds time for a sizable amount of utterly idle speculation. For instance, me: I've developed a theory about crows. It goes like this.

Crows are bored. They suffer from being too intelligent for their station in life. Respectable evolutionary success is simply not, for these brainy and complex birds, enough. They are dissatisfied with the narrow goals and horizons of that tired old Darwinian struggle. On the lookout for a new challenge. See them there, lined up conspiratorially along a fence rail or a high wire, shoulder to shoulder, alert, self-contained, missing nothing. Feeling discreetly thwarted. Waiting, like an ambitious understudy, for their break. Dolphins and whales and chimpanzees get all the fawning publicity, great fuss made over their near-human intelligence. But don't be fooled. Crows are not stupid. Far from it. They are merely underachievers. They are bored.

Most likely it runs in their genes, along with the black plumage and the talent for vocal mimicry. Crows belong to a remarkable family of birds known as the Corvidae, also including ravens, magpies, jackdaws, and jays, and the case file on this entire clan is so full of prodigious and quirky behavior that it cries out for interpretation not by an ornithologist but by a psychiatrist. Or, failing that, some ignoramus with a supple theory. Theoretical ecologists adept at computer modeling can give us those fancy equations depicting the whole course of a creature's life history in terms of energy allotment to every physical need, with variables for fertility and senility and hunger and motherly love, but they haven't yet programmed in a variable for boredom. No wonder the Corvidae dossier is still packed with unanswered questions.

At first glance, though, all is normal. Crows and their corvid relatives seem to lead an exemplary birdlike existence. The home life is stable and protective. Monogamy is the rule, and most mated pairs stay together until death. Courtship is elaborate, even rather tender, with the male doing a good bit of bowing and dancing and jiving, not to mention supplying his intended with food; eventually he offers the first scrap of nesting material as a sly hint that they get on with it. While she incubates a clutch of four to six eggs, he continues to furnish the groceries and stands watch nearby at night. Then, for a month after hatching, both parents dote on the young. Despite strenuous care, mortality among fledglings is routinely high, sometimes as high as 70 percent, but all this crib death is counterbalanced by the longevity of the adults. Twenty-year-old crows are not unusual, and one raven in captivity survived to age twenty-nine. Anyway, corvids show no inclination toward breeding themselves up to huge numbers, filling the countryside with their kind (like the late passenger pigeon or an infesting variety of insect) until conditions shift for the worse and a vast population collapses. Instead, crows and their relatives reproduce at roughly the same stringent rate through periods of bounty and austerity, maintaining levels of population that are modest but consistent and that can be supported throughout any foreseeable hard times. In this sense they are astute pessimists. One consequence of such modesty of demographic ambition is to leave them with excess time, and energy, not desperately required for survival.

The other thing they possess in excess is brain power. They have the largest cerebral hemispheres, relative to body size, of any avian family. On various intelligence tests—to measure learning facility, clock-reading skills, the ability to count—they have made other birds look doltish. One British authority, Sylvia Bruce Wilmore, pronounces them "quicker on the uptake" than certain well-thought-of mammals like the cat and the monkey, and admits that her own tamed crow so effectively dominated the other animals in her household that this bird "would even pick up the spaniel's leash and lead him around the garden!" Wilmore also adds cryptically: "Scientists at the University of Mississippi have been successful in getting the cooperation of Crows." But she fails to make clear whether that was as test subjects or on a consultative basis.

From other crow experts come the same sorts of anecdote. Crows hiding food in all manner of unlikely spots and relying on their uncanny memories, like adepts at the game of Concentration, to find the caches again later. Crows using twenty-three distinct forms of call to communicate various kinds of information to one another. Crows in flight dropping clams or walnuts on highway pavement, to break open the shells so the meats can be eaten. Then there's the one about the hooded crow, a species whose range includes Finland: "In this land Hoodies show great initiative during winter when men fish through holes in the ice. Fishermen leave baited lines in the water to catch fish and on their return they have found a Hoodie pulling in the line with its bill, and walking away from the hole, then putting down the line and walking back on it to stop it sliding, and pulling it again until [the crow] catches the fish on the end of the line." These birds are bright.

And probably—according to my theory—they are too bright for their own good. You know the pattern. Time on their hands. Underemployed and overqualified. Large amounts of potential just lying fallow. Peck up a little corn, knock back a few grasshoppers, carry a beakful of dead rabbit home for the kids, then fly over to sit on a fence rail with eight or ten cronies and watch some poor farmer sweat at the wheel of his tractor. An easy enough life, but is this *it*? Is this *all*?

If you don't believe me, just take my word for it: Crows are bored.

And so there arise, as recorded in the case file, these . . .—no, *symptoms* is too strong. Call them, rather, *patterns of gratuitous behavior*.

For example, they play a lot.

Animal play is a reasonably common phenomenon, at least among certain mammals, especially in the young of those species.

Play activities, by definition, are any that serve no immediate biological function and that therefore do not directly improve the animal's prospects for survival and reproduction. The corvids, according to expert testimony, are irrepressibly playful. In fact, they show the most complex play known in birds. Ravens play toss with themselves in the air, dropping and catching again a small twig. They lie on their backs and juggle objects (in one recorded case, a rubber ball) between beak and feet. They jostle one another sociably in a version of king of the mountain with no real territorial stakes. Crows are equally frivolous. They play a brand of rugby, wherein one crow picks up a pebble or a bit of shell and flies from tree to tree, taking a friendly bashing from his buddies until he drops the token. And they have a comedy-acrobatic routine: allowing themselves to tip backward dizzily from a wire perch, holding a loose grip so as to hang upside down; spreading out both wings, then daringly letting go with one foot; finally, switching feet to let go with the other. Such shameless hotdogging is usually performed for a small audience of other crows.

There is also an element of the practical jokester. Of the Indian house crow, Wilmore writes that "this Crow has a sense of humor, and revels in the discomfort caused by its playful tweaking at the tails of other birds, and at the ears of sleeping cows and dogs; it also pecks the toes of flying foxes as they hang sleeping in their roosts." This crow is a laff riot. Another of Wilmore's favorite species amuses itself, she says, by "dropping down on sleeping rabbits and rapping them over the skull or settling on drowsy cattle and startling them." What we have here is actually a distinct subcategory of playfulness known, where I come from at least, as cruisin' for a bruisin'. It has been clinically linked to boredom.

Further evidence: Crows are known to indulge in sunbathing. "When sunning at fairly high intensity," writes another British corvidist, "the bird usually positions itself sideways on to the sun and erects its feathers, especially those on head, belly, flanks and rump." So the truth is out: Under those sleek ebony feathers, they are tan. And of course sunbathing (like ice fishing, come to think of it) constitutes prima facie proof of a state of paralytic ennui.

But the final and most conclusive bit of data comes from a monograph by K.E.L. Simmons that appeared in the *Journal of Zoology*, published in London. (Perhaps it's for deep reasons of national character that the British lead the world in the study of crows; in England, boredom has great cachet.) Simmons's paper is curiously titled "Anting and the Problem of Self-Stimulation." *Anting* as used here is simply the verb (or, to be more precise, participial) form of the insect. In ornithological parlance, it means that a bird—for reasons that remain mysterious—has taken to rubbing itself with mouthfuls of squashed ants. Simmons writes: "True anting consists of highly stereotyped movements whereby the birds apply ants to their feathers or expose their plumage to the ants." Besides direct application, done with the beak, there is also a variant called *passive anting*: The bird intentionally squats on a disturbed anthill, inviting hundreds of ants to swarm over its body.

Altogether strange behavior, this antic anting—and especially notorious for it are the corvids. Crows avidly rub their bodies with squashed ants. They wallow amid busy ant colonies and let themselves become acrawl. They revel in formication.

Why? One theory is that the formic acid produced (as a defense chemical) by some ants is useful for conditioning feathers and ridding the birds of external parasites. But Simmons cites several other researchers who have independently reached a different conclusion. One of these scientists declared that the purpose of anting "is the stimulation and soothing of the body," and that the general effect "is similar to that gained by humanity from the use of external stimulants, soothing ointments, counter-irritants (including formic acid) and perhaps also smoking." Another compared anting to "the human habits of smoking and drug-taking" and maintained that "it has no biological purpose but is indulged in for its own sake, for the feeling of well-being and ecstasy it induces."

You know the pattern. High intelligence, large promise. Early success without great effort. Then a certain loss of purposefulness. Manifestations of detachment and cruel humor. Boredom. Goofing off. Anomie. And finally . . . the dangerous spiral into drug abuse.

But maybe it's not too late for the corvids. Keep that in mind next time you run into a raven, or a magpie, or a crow. Look the bird in the eye. Consider its frustrations. Try to say something stimulating.

The Widow Knows

HERE'S A CHEERFUL THOUGHT: Some knowledgeable people believe that black widow spiders, like locusts and jackrabbits, come in plagues.

No one, of course, has kept a precise running count on our total supply. Since the black widow spider is by nature shy, almost fanatically discreet, the intermittent explosions of black widow populations are gauged by extrapolation from the number of bites suffered by humans. At certain historical junctures of place and time, those widow bites have reached what are called "epidemic" levels. Spain had such an epidemic in 1830. Eastern Russia had another in 1838–39. France endured a peak forty years after Russia, Uruguay in 1910, and then, in 1926, black widows terrorized Yugoslavia. These outbreaks of spider bite and spider fear were all presumably caused by members of the genus *Latrodectus*, not precisely the same species as our American black widow but closely related. The last major episode of notoriety for *Latrodectus mactans*, our domestic version, occurred a half-century ago. In the autumn of 1934, with Huey Long gamboling in the Senate and John Dillinger freshly slain on the streets of Chicago, Americans were suddenly worried about black widow spiders.

The widow boom that year was no doubt a combined result of climatic conditions, ecological cycles, and publicity. Through the mild winter and dry summer of 1933, more black widows had been surviving, raising larger broods, biting more humans, and getting more journalistic attention for doing so. The Associated Press carried continuing reports on the condition of two unfortunate men, one in Alabama and one in Idaho, both dangerously ill from black widow bites. Local newspapers ran stories about barbarous mortal battles staged between black widows and scorpions, black widows and tarantulas, even black widows and snakes. With a peculiar repulsive fashionableness, black widows were *in*. Then *Scientific American* fanned the coals with an article announcing in careful detail why *Latrodectus mactans* should rightly be feared, and in November the sober journal *Science* published a note entitled "On the Great Abundance of the Black Widow Spider."

The note's author speculated about effects of the recent weather, commented that cousins and competitors of the widow seemed to be in eclipse, and concluded, "Possibly *L. mactans* is beginning to get the upper hand in the great struggle for existence." But the struggle against whom? To some nervous observers, this upper-handhood seemed ominous in a creature with eight legs, not to mention a pair of poisonous chops, so within months that statement from *Science* had been translated—now in the less sober pages of *Popular Mechanics*—into this one: "Aided by favorable climatic conditions, the spider has multiplied so rapidly that it is becoming a real menace to man." The headline in *Popular Mechanics* was "Wasps May End Black Widow Spider Menace." Entomologists were groping about for some spider-eating savior, a natural predator, a nemesis, to turn back those menacing, swelling, surging hordes of *Latrodectus mactans*.

Two species of wasp were considered, a mud dauber that kills adult black widows as food for its young and another form that chews its way into the silken egg sac and deposits its own eggs where, after hatching, the wasp larvae can eat the unborn spiders. There was also talk that black widows might be subdued by a species of small parasitic fly, or by alligator lizards allowed to run rampant, or by a certain brave sort of toad, or perhaps by spider-eating spiders of the genus *Mimetus*. People were desperate.

Yet these scientists and other widow-watchers needn't have been quite so concerned.

The Great Widow Scare of 1934 came and went, like the King-fish and Dillinger, but no airplanes were ever called out to spray. No frantic eradication campaign or mass deployment of wasps ever took place. The crisis—more accurately, the perception of a crisis—passed away inconclusively, thanks to some form of natural equilibration. *Latrodectus mactans* continues to be widely distributed, present in every American state and common in many, abundant in most of the warmer ones. Chances are good that wherever you live, during the last year sometime you have sat down within ten feet of a black widow. (If you had occasion to use an old wooden outhouse, among the widow's favorite habitats, you might even have been cheek-by-jowl with one.) Still, the species has never become a large-scale problem. Being bitten is nothing to take lightly, and if the victim is an old person or a small child, it can be fatal. But this is quite rare. The plague hasn't happened. The widow hordes, with their eight upper hands in the great struggle for existence, have never come marching like driver ants over the hill into your town or mine. Why not? Partly because, in an ecological sense, the black widow knows its place.

More specifically, *she* knows her place. The most efficient natural predator controlling any population of black widow spiders is very likely the female black widow. In that delicate Malthusian balance between vast reproductive potential (many eggs per sac) and limited life-supporting resources (only so many good spots to build a web, and only so much food available to be netted), the female of the species serves a gyroscopic function.

Concerning the behavior of the female of *Latrodectus mactans*, one point should be clarified. A black widow spider is not like a mad dog or a vicious human: She doesn't bite without reason, she doesn't kill without purpose. She is not necrophilic or otherwise kinky. Yes, sometimes she does lasso her chosen male after copulation, as he makes for the exit, and suck him dry as a roasted chili. But that only happens if she's hungry.

The female widow's hunger, or lack of it, is the standard against which certain life-or-death decisions are made, and those decisions exert a geometrically multiplied impact on overall population levels of the species. When the concentration of widows in any area is high, competition is fierce for good web sites and ambient provender. As a result, each widow tends to be hungrier. The female has an amazing capacity to endure long stretches without food—three or four months, at least, and one widow in captivity went unfed for nine months, then was nursed back to health— which makes it unlikely that the species could ever die out entirely from starvation. But as competition grows keen and food grows scarce, the female widow takes some drastic population-control measures that tip the balance back toward circumstances of lonely affluence. In the process, she also slakes her own hunger. The first of these measures involves mating.

The male widow, much smaller than the female and more mobile across unfamiliar terrain, takes the romantic initiative. He appears at the edge of the female's web, sets his feet on a few strands, and, by bobbing his abdomen, causes the whole web to vibrate. This is the mating offer. If the female is not in condition to breed or not in the mood, she will not respond. But if she happens also to be very hungry, she may wait cryptically for him to venture within reach, then grab him, swathe him in silk, and eat him. If she *is* ready to mate, there is an answering pattern of web vibrations from her. After about two hours of foreplay, during which he wraps her in a loose veil of silk, they consummate. That takes about five minutes, but the positions are exotic.

As they uncouple, the question again arises: Is she hungry? If so, he is snatched back and devoured. If not, he is allowed to leave peaceably, or he may linger in a corner of her web, safely ensconced as mate emeritus, until accelerated senescence kills him a few days later. Whichever of these outcomes occurs, the female has remained in character, neither sentimental nor sadistic but merely practical, all eight of her eyes fixed on the basics of survival. She is the incarnate future of her species, far more so than the male, and in a deeply instinctive way she acts upon that responsibility.

The same sort of chilly pragmatism governs her maternal behavior. She's a solicitous mother, but only up to a certain point. Her eggs are laid—anywhere from 25 to 1,000 in a batch, on average about 200—and then wrapped by her in a silken cocoon. This egg sac is watertight and reflective; it shields the eggs from sunlight, rain, and predators and helps keep them warmly incubated. The mother has taken great care over it. She will fight to protect it. She will spend immense effort moving it from one part of the web to another, sunny exposure to shade, for the sake of maintaining its optimal temperature. Then, when the eggs hatch and her young crawl out, if the message of hunger has by now again reached her and the prospects seem meager for black widows, she will eat them.

Or at least some of them. After thirty or forty she may stop, and the rest of the spiderlings will go off to face all the other obstacles, including starvation, between them and adulthood.

The behavioral ecology of *Latrodectus mactans* has been studied that far, but not much farther. Even the scientists, even the *arachnologists*, are still puzzled; and their puzzlement is our ignorance. Does the maternal black widow perceive some conscious or unconscious signal about how many young spiders might be expected to survive? We don't know. Is the male widow sexually capable of servicing more than one female? We don't know. Does the possibility of his fathering further broods influence his first mate in her choice of whether or not to kill him? We don't know. And the female, with her edgy perceptions of hunger, her hair-trigger readiness to cannibalize—has she been programmed by evolution to damper fluctuations in the population of her species? It seems plausible. But we don't know.

There's a welter of uncertainties. If the female black widow wore a red question mark on her abdomen instead of an hourglass, she could scarcely better exemplify the limitations of our understanding of this—and many other—familiar but neglected aspects of the natural world. Who knows what graceful intricacies lurk in the behavioral wiring of *Latrodectus mactans*? Not us. Maybe that's why she seems so spooky, so necessary, and so beautiful.

The Troubled Gaze
of the Octopus

IN *GRAVITY'S RAINBOW,* Thomas Pynchon's great steaming slag heap of a novel, there is a memorable scene in which an enormous octopus comes slouching out of the sea, grabs a young woman around the waist with one sucker-studded arm, and tries to drag her back into the water. Echoes of King Kong and Fay Wray. The woman is rescued by Pynchon's hero, a certain Tyrone Slothrop, who pummels the octopus over the head with a wine bottle, to no effect, and then distracts it by offering a tasty crab. "In their brief time together Slothrop forms the impression that this octopus is not in good mental health," Pynchon tells us. I have sometimes had the same feeling about Thomas Pynchon, but never mind. The point of recalling that octopus scene in particular is that I always took it to be luridly and outlandishly surreal, another hallucinatory cartoon caper, like so much that comes out of Mr. Pynchon's rich and febrile imagination.

It turns out, though, that I was unduly skeptical. It turns out that this scene might contain more truth than poetry. The details of anatomy and scale and behavior have a sound basis in zoological reality; only the matter of motivation remains unsettled. The species in question is *Octopus dofleini*, otherwise known as the giant Pacific octopus.

Does this creature really lay hold, in such peremptory manner, of unsuspecting human beachgoers? And if so, just what has it got in mind?

The concept of *mind* is not inappropriate as applied to the octopuses, since these creatures have by far the most highly developed brain in their province of the animal kingdom. They belong to the phylum Mollusca, a large group of invertebrates mainly characterized by soft bodies, hard shells, and rather primitive patterns of anatomical organization, well suited to surviving inconspicuously on the sea bottom. Typical of the Mollusca are clams, oysters, snails; the octopuses (and to a lesser extent their near relatives, the squids) are decidedly untypical. They are an evolutionary anomaly, a class of hyperintelligent misfits who have advanced far beyond their origins.

The octopuses have an elaborate fourteen-lobed brain, an organ so large that their brain-to-body ratio exceeds that for most fishes and reptiles. Mentally, they are more on a level with birds and mammals. They possess a capacity for learning, memory, and considered behavior that makes them the smartest of all sea-dwelling animals, with the exception of marine mammals. In a laboratory, they tend to be good at mazes and perform well in tests of discrimination among visual symbols. This last talent depends partly on their acute eyesight. Every octopus looks out at the world through a pair of extraordinary eyes—eyes about which, to a human, there is something unexpectedly and disquietingly familiar.

"The animal has eyes that stare back," according to Martin J. Wells, a British zoologist who is one of the world's experts in octopus physiology and behavior. "It responds to movement, cowering if anything large approaches it, or leaning forward in an alert and interested manner to examine small happenings in its visual field." Jacques Cousteau goes a bit further: "When a diver sees a giant octopus in the dim water, its great eyes fixed on him, he feels a strange sensation of respect, as though he were in the presence of a very wise and very old animal, whose tranquillity it would be best not to disturb." One of Cousteau's assistants adds: "I have often had the impression that they are 'reflecting.'" Other divers and lab researchers make the same sort of comment, describing the same eerie sense of encounter, recognition, even mutuality. Lately I've had occasion to experience it myself, during three evenings of octopus-watching in a small university room filled with quietly gurgling tanks: the potent, expressive gaze of the octopus. These animals don't just gape at you glassily, like a walleye. They make eye contact, as though they are someone you should know.

One reason for the potency of that stare is simply a matter of proportion. Relative to the body size of a given octopus, the eyes are, like the brain, unusually large. (The ultimate record in this regard belongs to that octopus cousin the giant squid, with an eyeball up to fifteen inches across, the largest on Earth and twice the size of the eye of a blue whale.) Octopus eyes are also protrusive and mobile, bulging up periscopically when the creature's attention is caught, swiveling far enough fore and aft to cover all 360 degrees of horizon. But the real magic behind the octopod gaze is that those eyes bear a startling structural similarity to our own.

It's an exemplary instance of the phenomenon called convergent evolution. Two separate evolutionary paths are followed for millions of years by two disparate groups of creatures, arriving eventually at two separate but (coincidentally) very similar solutions to a common problem. In this case, the problem of translating incident light rays into coherent images conveyable to the brain. The vertebrate eye—the model we humans share with cougars and eagles and rattlesnakes, all having inherited the pattern commonly—is an ingenious contrivance combining a cornea, a crystalline lens, an adjustable iris, and a retina. That such an organ evolved even once, within the vertebrate line, represents a remarkable triumph of time and trial-and-error over improbability. The still more improbable happenstance is that *two* very similar versions of this piece of biological engineering have appeared independently. The other belongs exclusively to the octopods and their close kin. Each of those squid and octopus eyes consists of a cornea, a crystalline lens, an adjustable iris, and a retina, functioning together in much the same way as ours.

And the octopuses are also endowed with an eyelid, so that they can wink at us fraternally.

Among all the sea's octopod eyeballs, the most imposing belong to *Octopus dofleini*. This is the giant that has impressed Cousteau and others with its dignified presence, as though it were "a very wise and very old animal." The wisdom part is quite possible, but the great agedness is an illusion.

Octopuses grow quickly to adulthood and die at an early age, in most cases right after their first breeding experience. Two or three years is a full life span, even for the larger Mediterranean octopuses, which might grow in that brief time from the size of a flea to the size of an ottoman. An octopus can achieve such speedy growth—almost doubling its weight each month throughout most of its life—because of its exceptional metabolic efficiency in converting food protein to octopus protein. And at this process *O. dofleini* is probably unsurpassed. Surviving longer than other species, to the grand age of five or maybe a little beyond, some giant Pacific octopuses attain awesome sizes.

They live mainly in sea-bottom caves along the coast of the northern Pacific, from California up through British Columbia and Alaska and across to Japan. The caves give them security from predation and, at mating time, a good place to brood eggs. They seem to prefer the range of moderate depths from the intertidal zone down to a hundred fathoms, and by most accounts they are exceedingly shy. Until three or four decades ago, *O. dofleini* were known almost solely from commercial fishermen, who occasionally, inadvertently and probably to their own vast alarm, brought one up in a trawl net. After World War II, improved diving equipment (especially scuba) opened a new degree of human access to those deeper caves. Reports of larger and larger *dofleini* began to appear. Cousteau mentions one specimen, spotted off Seattle, with an arm span of 30 feet and a weight around 200 pounds. Another diver has told William High, of the National Marine Fisheries Service, about bringing up several 400-pound *dofleini* during his commercial octopus-fishing days, as well as a single huge individual that went 600 pounds.

Then the inevitable happened. Someone had a clever idea, and in 1956 hundreds of divers converged on Puget Sound to compete in an event billed as the World Octopus-Wrestling Championship. It became an annual tradition.

The biggest specimens of *dofleini* were smoked out of their caves with solutions of noxious chemicals and wrestled up onto land by divers working in teams, there to be weighed and measured and admired. Dealing underwater with an octopus that large requires—from a human in scuba gear—equal measures of skill, coolheadedness, and lunatic daring. *Dofleini* are naturally timid but also quite strong, and they do have four times as many arms. Panic-stricken *dofleini* have been known to pin a man's arms to his sides, pull off his face mask, yank out his mouthpiece. There have even been several instances when a big octopus pounced on the back of an unsuspecting diver from a rock ledge overhead. If one diver becomes trapped, in a situation like that, the usual procedure seems to be for his buddy to commence slicing away octopod arms with a knife or go for the eye with a spear.

After those "wrestling" championships each year, the healthy octopuses were carefully released back into the sea. No harm done. No permanent toll on the giant octopus population. Right? At least, that was the assumption. But one woman diver who took part in the roundups has told Cousteau: "It is hard to keep in mind that octopuses of the size and weight of these are really very fragile animals, highly developed and with a very sensitive nervous system. They seem to succumb easily to nervous disorders. If a diver is too rough with an octopus, even without actually hurting it physically, it happens that the animal goes into a state of emotional shock and sometimes dies."

Let me recapitulate. It seems that 1) octopuses in the 200-pound range, or larger, cowering in sea caves off the coast near Seattle, and 2) known for their high-strung susceptibility to nervous disorders, have been 3) kidnapped and terrorized intermittently by strange visitors in black neoprene, these latter often armed with knives and spears. Consequently it can be assumed that 4) eyeballs humanlike but the size of grapefruit now gaze out from those caves, furtively, trepidatiously, some of them no doubt looking just a bit addled—looking, that is, as though they might belong to animals that, like Pynchon's beast, are in not quite the best mental health.

My own view is this: Any giant octopus that grabs hold of a passing human probably has some pretty good reason. If not an unanswerable grievance, then at least a plausible insanity defense.

Or maybe the grabby creature is just desperate to communicate. Snatching that startled human by the neoprene lapels; exigent as the Ancient Mariner; transfixing the man or woman with a big glittering eye. As though to say: *Listen. We know who you are. And we've seen what you do.* But unfortunately the octopuses, for all their intelligence, for all their sensitivity, for all their evolutionary sophistication, are born mute.

Avatars of the Soul in Malaya

CONSIDER NOW THE LEPIDOPTERA, in all their vacant splendor.

They are the bimbos of the natural world: more beautiful and less interesting, arguably, than any other order of animals. An evolutionary experiment in sheer decorative excess, with a high ratio of surface to innards. They move through the air like pulses of idle thought. They have a weakness for flowers. They are prodigiously diverse without being adventurous: roughly 150,000 known species, all of which behave pretty much alike; 150,000 distinct patterns, but in each case a six-legged worm strung between kites. They are silent. Detached and diaphanous. Generally they possess neither teeth nor jaws. They feed pacifically on plant liquids or (some species) just go hungry through their entire adulthood. Fly on wings that are fleshless and papery, flashing bright iridescent colors produced by the devious exploitation of tiny prisms and mirrors. Certainly these are real physical creatures, yes; then again, they just don't seem to be quite all *there*. Aristotle was onto something, I think, when in the fourth century B.C. he used the Greek word *psyche* to mean both "soul" and "butterfly."

They might be insects. Or they might be platonic ideas.

In classical Greece and then later in Rome, this link with the spiritual realm was applied to both major subgroups of Lepidoptera, the moths as well as the butterflies. Both moths and butterflies were delicate enough to suggest a pure being, freed of its carnal envelope. Both were known to perform a magical metamor-phosis—from fat, ugly caterpillar to gorgeous, airborne adult, with a dormant pupal stage in between—that put humans in mind of resurrection from the grave. Moths may have been even more suited than butterflies to bearing this burden of symbolism, in that moths, like ghosts, fly at night. Tomb-sculpture designs from impe-rial Rome have survived (thanks to later Italian scholars, who copied them before the original stones were lost) on which appear butterflies and moths carved to represent the departing souls of the dead. And the motif has endured. One marker from a nineteenth-century grave in Massachusetts shows a common mon-arch, *Danaus plexippus*, freshly emerged from its chrysalis and winging away—the soul as butterfly.

Clearly mankind has taken some small comfort over the past couple millennia from gazing upon Lepidoptera and positing this odd connection between their substance and our essence. Why the Lepidoptera? Maybe it's *because* they are detached and diapha-nous, *because* their beauty is of an otherworldly sort. No sting, no bite, no bothersome buzz. Strict vegetarians. They represent an ideal of sweetness and gentle grace that seems almost innocent of the whole merciless evolutionary free-for-all. No wonder they are, zoologically, so godawful boring to contemplate.

But wait. That's only the traditional, happy-time view of the Lepidoptera. It applies to no more than about 149,999 species. And it takes no account whatever of a small Malayan jungle moth called *Calpe eustrigata*.

Here finally is a moth with character, a moth with edge, a moth unafraid to besmirch itself in the Darwinian struggle. A moth unique, among all known moths, for its dietary behavior. *Calpe eustrigata* sucks blood from humans.

There is no common name for *Calpe eustrigata*, but it belongs to a large family of drab little moths called the noctuids, notable in this country mainly for the damage that larvae of some species do to vegetables and grain. These are your basic cutworms and ear-worms and celery-loopers. Such dreary noctuids can be plentiful, species by species, but *C. eustrigata* is quite rare. It was discovered some decades ago by an admirable fanatic named Hans Bänziger, a Swiss entomologist who was spending two years in the jungles of Thailand for research on a different group of noctuids.

In Thailand the work progressed satisfactorily; the moths behaved more or less as expected, and there was no sign of any such creature as *C. eustrigata*. Toward the end of two years, Bänziger got down into Malaya (as peninsular Malaysia was then called), where he wanted to investigate several species that were opportunistic blood-drinkers of a purely nonaggressive sort. These Malayan moths were known to lick at the open wounds of large mammals and to follow after mosquitoes (which are greedy and slovenly as they extract blood), lapping up what was spilled. Again the moths behaved as expected. At least they did until, late one night, Bänziger captured a particular specimen. He found it alighted on a water buffalo.

Bänziger's own account, from a back issue of the journal *Fauna*, sets the scene: "I had become suspicious of this insect species because of a photograph taken a few days before while it was feeding on a Malay tapir. The photograph showed something very strange about the moth's proboscis. Now with a live specimen I intended to study its feeding behavior on myself. That night was to become especially exciting! Having incised my finger with a scalpel to draw fresh blood, I offered my finger to the caged moth. The moth climbed onto my finger and did in fact plunge its proboscis into the blood, but it appeared to imbibe none. Instead it stuck its straight, lancelike proboscis into the wound and, without any regard for the donor, penetrated the flesh. The pain I felt caused me to utter a cry of—joy!" Lucky the man who so loves his work. "I had discovered a moth which *pierces* to obtain blood." If you were a lepidopterist, you'd see that it was a pretty big moment in history.

During a month in Malaya, Bänziger found only twenty-four more specimens of *C. eustrigata*, but he kept them busy poking and sucking at his hand. As an experimentalist, he lacked nothing by way of commitment. "Blood-sucking occupied from 10 to 60 minutes and blood continued to flow out of the wound for a few minutes after cessation. Hours and sometimes even days later, the wound was still itching." The remarkable aspects of all this involved not just the matter of behavior (including Bänziger's behavior) but the matter of anatomy. Lepidopteran mouth structures are totally different from those of, say, a mosquito. The standard equipment for butterflies and moths is a long flexible tube that remains coiled up under the head until, when needed, it can be sprung out straight by hydraulic pressure, like one of those paper noisemakers in the mouth of a drunk on New Year's Eve. In extended position it allows the insect to suck nectar from the reservoir of a deep flower. But this thing is a drinking straw, not a drill. Until Bänziger, no lepidopterist had ever seen a moth whose proboscis could be stabbed through human flesh.

There was also the mystery of its evolution. Where had *C. eustrigata* come from? How had the bloodsucking adaptation, along with the equipment to practice it, arisen? What were the intermediate stages between Bänziger's new species and those other Lepidoptera—all 149,999 of them—who noodle from blossom to blossom drinking nectar through their elongated schnosters? What manner of temptation could have lured certain moth species astray, turning their tastes from flowers to blood?

The answer, it seems, was fruit. Faced with mortal competition over limited supplies of nectar, a number of noctuid species have adapted themselves to feeding on the juices of overripe fruit. Some have developed stronger and sharper mouth tubes that allow them to pierce the skin of soft fruits, such as peaches and raspberries, and suck out their fill of juice. One species is even armed with a proboscis that will penetrate the skin of an orange. Among these fruit-piercing moths are several close cousins of *C. eustrigata*. Bänziger suggests that in a habitat where fruit was available only seasonally but juicy mammals were present year-round, desperate necessity might have led to the next logical step: vampirism, as practiced by moths.

But the vampire moth was just a distraction from what had taken Hans Bänziger out to Southeast Asia. He was there to study a group of species he called, rather blandly, the "eye-frequenting" Lepidoptera—moths that literally live on a diet of tears.

These wondrous creatures may have evolved from the opportunistic blood-drinkers that clean up after messy mosquitoes. Bänziger says: "Probably by crawling about on their mammalian hosts some moths found the eyes, where there are always discharges. And thus there evolved the habit of dining exclusively upon eye discharges, which contain various proteins such as globulin, albumen, and others in the leucocytes and epithelial cells in tears." Such moths, in the wild, would drink from the eyes of elephants. They would drink from the eyes of horses, buffalo, antelope, or pigs. And as Bänziger demonstrated with his characteristic élan, given a proper opportunity, they would drink from the eyes of man. "The lachrymal secretion was very much stimulated by the activity of the moth. After 30 min. my eye was so irritated that I was forced to interrupt the experiment." Only thirty minutes of moth-in-eye jabbing, imagine, and he called a halt. But not before the insect had drunk freely from Bänziger's own tears of—joy! There is even a photograph showing a small noctuid of the species *Lobocraspis griseifusa* perched head-down across Bänziger's brow, its tube extended to drink fluid off the surface of his cornea. "Note the deep penetration," reads his unblinking caption, "of the proboscis between eye and eye lid."

Note the deep penetration of Lepidoptera between fact and imagination. I suspect that the Greeks and the Romans would have known what to make of *Lobocraspis griseifusa*, a species in the spiritual tradition. Imagine how Ovid would have loved those insects. The souls of the dead return, on powdered wings and in silence, to comfort mankind; to console us earthbound survivors; to drink away our very tears

It's a nice thought—too nice to be true. But the intricate realities of evolutionary entomology don't lend themselves so well (no better than other branches of nature) to our neat romances and our pathetic fallacies, and these two species stand as vouchers to that truth. If we idealize the tear-drinkers, then what about the suckers of blood? A moth, hungry animal, is only what it is.

Rumors of a Snake

WHAT THE WORLD NEEDS IS A GIANT SNAKE. Make it vicious, make it loathsome, make it sixty feet long—and let it swim in the waters of the Amazon.

All right, don't look at me, this isn't my personal obsession. Apparently there exists a popular imperative to that effect, a deep-seated hope or dark desire, issuing from the delicately balanced ecosystem of collective human consciousness. *Give us a huge snake. A serpentine monster at the far fringe of imaginability. Let it inhabit the wettest mires of the deepest jungle. A horrific thing, slithering along in elegant menace, belly distended with pigs and missing children.* The evidence for this weird yearning is oblique but cogent: Lacking any such beast, we are eager to settle for rumors of one. Otherwise how to explain the breathless compoundment of hearsay, tall tale, and hyperbole that has always surrounded *Eunectes murinus*, an honest and tangible animal commonly called the anaconda?

One fair example appears in the memoirs of Colonel Percy Fawcett, a British archeologist and explorer of the early twentieth century. In 1906, Fawcett was sent out from London by the Royal Geographical Society to make a survey along certain rivers in western Brazil. "The manager at Yorongas told me he killed an anaconda 58 feet long in the Lower Amazon. I was inclined to look on this as an exaggeration at the time, but later, as I shall tell, we shot one even larger than that." The disclaimer about "I was inclined . . ." is a cagey stroke. Major Fawcett would have us take him for a hardheaded British skeptic.

Later he tells: "We were drifting easily along in the sluggish current not far below the confluence of the Rio Negro when almost under the bow of the *igarité* there appeared a triangular head and several feet of undulating body. It was a giant anaconda. I sprang for my rifle as the creature began to make its way up the bank, and hardly waiting to aim smashed a .44 soft-nosed bullet into its spine, ten feet below the wicked head. . . . We stepped ashore and approached the reptile with caution. It was out of action, but shivers ran up and down the body like puffs of wind on a mountain tarn. As far as it was possible to measure, a length of forty-five feet lay out of the water, and seventeen feet in it, making a total length of sixty-two feet." The indisputable logic of good arithmetic. It might all be true but most likely it isn't.

An adventurer of the 1920s named F. W. Up de Graff offered a similar account, having observed his anaconda in shallow water: "It measured fifty feet for certainty, and probably nearer sixty. This I know from the position in which it lay. Our canoe was a twenty-four footer; the snake's head was ten or twelve feet beyond the bow; its tail was a good four feet beyond the stern; the center of its body was looped up into a huge S, whose length was the length of our dugout and whose breadth was a good five feet." But size estimates made in a watery medium are notoriously unreliable—especially when that watery medium is the Amazon River.

Bernard Heuvelmans, in his feverish book of cryptozoology, *On the Track of Unknown Animals*, tells at third or fourth hand of another Brazilian specimen that was purportedly killed in 1948: "The snake, which was said to measure 115 feet in length, crawled ashore and hid in the old fortifications of Fort Tabatinga on the River Oiapoc in the Guaporé territory. It needed 500 machine-gun bullets to put paid to it. The speed with which bodies decompose in the tropics and the fact that its skin was of no commercial value may explain why it was pushed back in the stream at once." Always for the gigantic individuals there is this absence of physical evidence, and always a waterproof reason for the absence: no cameras on hand, rotting meat, even the skin was too heavy to carry out. One photograph did exist that, during the 1950s, was sold all over Brazil as a postcard, its caption claiming a length of 131 feet for the snake pictured. Unfortunately, no object of reference appeared in the photo with it. That snake might as easily have been a robust but minuscule 20-footer.

My own modest sighting comes not from Brazil but from northeastern Ecuador, along the Rio Aguarico, in a remote zone of lowland jungle that may be as favorable to the production and growth of anacondas as almost anywhere in the Amazon drainage. Like those other wide-eyed witnesses Fawcett and Up de Graff, I was in a dugout canoe, along with a dozen or so adventuresome tourists. Our guide was an intrepid and jungle-smart young man named Randy Borman, raised there in the forest among Cofane Indians. Randy spotted the big snake on a log tangle near the riverbank while the rest of us were gawking elsewhere. He steered the boat in for a closer look.

Dark gunmetal gray with sides mottled in reddish brown, the anaconda was sunning itself placidly. Barely above the water on a low-riding log; protectively colored and patterned so that even from ten yards away it was virtually invisible. Randy edged the canoe closer. Do you see it now? Some of us did; several admitted they didn't. We moved closer. Here indeed was a formidable snake. Still motionless, still sunbathing dreamily. I was delighted with this glimpse of an anaconda in the wild, but—amid the soft brushfire crackle of camera shutters—we had already ventured closer than I ever expected to get. Then closer still. A very tolerant and self-possessed snake. Beautiful big head. Thick graceful coils of body. Sizable brown eye. Hello, Randy? Just when I thought our guide would back the canoe off, instead he dove over the gunnel to grab this creature around the neck.

It was a deeply startling act. But Randy came up again, deftly, with a great armload of anaconda wrapping itself onto him in surprise and anger, squeezing with the authority of a species that does its killing by constriction. My face bore the contemplative expression of two eggs sunnyside in a white Teflon skillet.

Randy smiled calmly. "We'll take him back to camp for the others to see." An hour later, having been much fawned over and photographed, the animal was gently released back into its river. A few fast pulses of undulant swimming, then a dive beneath the brown water, and it was gone. There was a convenient absence of physical evidence.

So in recounting the story afterward (which I have not hesitated to do often, cornering people at parties and hoping the talk might turn to giant reptiles), I could make that poor snake any damn size I pleased. A piddling ten feet? Maybe eleven? Roughly the same girth as a man's biceps? In fact (so I would say, with coy dismissiveness) it was rather dainty as this species goes. Still, an impressive beast.

Very little is known about the biology of *Eunectes murinus*, the anaconda; even less about its life history in the wild; and a sad fact is that no one seems much to care.* Not a single field-research project, one expert has told me, is currently being done on it. The species has not yet found its George Schaller, its Dian Fossey, its Jane Goodall.

We know that it is a nonvenomous constrictor of the boa family. We know that it is aquatic, preferring slow rivers and swamps. That it bears live young (as opposed to laying eggs) in litters of up to eighty. That it is native to tropical South America east of the Andes, and also to the island of Trinidad. We know that (unlike other boas and pythons) it does poorly and often dies soon in cap-

---

* More is known, however, and more biologists seem to care, than when this essay first appeared in 1984. Field studies have indeed been conducted, yielding some interesting results. The herpetologist Harry Greene informs me that four distinct species of anaconda are now identified in South America. They differ in size and geographical distribution, among other ways. Two are widely distributed and somewhat familiar to science: our *Eunectes murinus*, today commonly called the green anaconda, which lives mainly in the Amazon basin and northern South America; and *Eunectes notaeus*, the yellow anaconda, which is considerably smaller at its average adult size (a mere 10 feet) and inhabits the swamps and rivers of southern Bolivia, Paraguay, and thereabouts. The other two species (*Eunectes deschauenseei* and *Eunectes beniensis*) are more narrowly distributed and still poorly known. Things have changed much in the realms of science and communication, not just since Percy Fawcett explored the Amazon but since I visited with Randy Borman. When further taxonomic and distributional information about anacondas is available, it will probably be posted on Wikipedia within hours.

tivity. But as to the rest, its favored diet, its daily and seasonal rhythms, its mating and birthing behavior, physiology, growth rate, longevity: almost a total blank. There is an absence of evidence.

Admittedly, the prospect of studying full-grown anacondas in their own habitat offers an array of uniquely forbidding logistical problems. For that reason or whatever others, scientific consideration of *Eunectes murinus* has been limited almost entirely to the same simple question that so mesmerized those early explorers: *How big does it get?* Well, really quite big. Bigger than any other snake on Earth. *But how big is that?*

A second sad fact about the anaconda: By scientific standards of verification, it just doesn't seem to be nearly so large as everyone seems to want to believe it is. Forget 131 feet. Forget 62 feet, even with faultless arithmetic. Discount the record-length skins, which generally have been stretched by a good 20 percent in the process of tanning. Scientists have their own unstretchable views on this matter.

One respected herpetologist, Afrânio do Amaral, has posited a maximum length for the anaconda of about 42 feet. But then Afrânio do Amaral is a Brazilian, arguably with a vested patriotic interest. And after him the figures only get stingier. James A. Oliver of the American Museum of Natural History was willing to grant 37½ feet, based on the measurement made by a petroleum geologist, with a surveyor's tape, of a snake shot along the Orinoco River. But again in this case there was the problem of physical evidence: "When they returned to skin it, the reptile was gone," we are told. "Evidently it had recovered enough to crawl away." Teddy Roosevelt is said to have offered $5,000 for a skin or skeleton 30 feet long, and the money was never claimed.* Sherman and Madge Minton, the authors of several reliable snake books, declare: "To the best of our knowledge, no anaconda over twenty-five feet long has ever reached a zoo or museum in the United States or Europe." And Raymond Ditmars, an eminent snake man at the Bronx Zoo early in this century, wouldn't believe anything over 19 feet.

---

* And some things *don't* change. The cash prize offered is now up to $30,000, Harry Greene tells me, and it still hasn't been claimed as of January 2007.

Can these people all be discussing the same animal? Can Dit-mars's parsimonious 19 feet be reconciled with the eyewitness account of Major Fawcett? Does Roosevelt's unclaimed cash square with Heuvelmans's 115 feet of worthless decaying snakeflesh? It seems impossible.

But new evidence has lately reached me that suggests an explanation for everything. The evidence is a small color photograph. The explanation is relativity.

Not Einstein's variety, but a similar sort, which I'll call Amazonian relativity. It's very simple: The true genuine size of an anaconda (this theory applies also to piranha and bird-eating spiders) is relative to three other factors: 1) whether or not the snake is alive; 2) how close you yourself are to it; and 3) how close both of you are, at that particular moment, to the Amazon heartland. A live snake is always bigger than a dead one, even allowing for posthumous stretch. And as the other two distances decrease— from you to the snake, and from you to the Amazon—the snake varies inversely toward humongousness.

This small color photograph, of such crucial scientific significance, arrived in the mail from an affable Dutch-born engineer, a good fellow I met on that Rio Aguarico trip. Unlike me, he had carried a camera, and sending the print was meant as a favor. In the photo's foreground I can see the outline of my own dopey duck-billed hat. The background is a solid wall of green jungle. At the center of focus is Randy Borman, astride the stern of his dugout, holding an anaconda. Dark gunmetal gray with sides mottled in reddish brown.

The snake is almost as big around as his wrist. It might be five feet long. Possibly close to six. But photographs can be faked. I don't believe this one for a minute.

Wool of Bat

FROM PLINY TO SHAKESPEARE to Tom McGuane, there has been a consensus: Any creatures so grotesquely improbable as bats must perforce lend themselves to some grotesquely improbable human use. The logic may be dubious, but the notion is long-standing.

During the first century A.D., for instance, Gaius Plinius Secundus (that is, Pliny the Elder) suggested in his *Natural History* that a drop of bat's blood hidden beneath a woman's pillow would affect her as an aphrodisiac. Twelve centuries later, Albertus Magnus claimed that smearing bat blood over your face like Coppertone would improve night vision. Macbeth's three witches, of course, are responsible for that famous stew recipe of which the partial ingredients are

> *Eye of newt, and toe of frog,*
> *Wool of bat, and tongue of dog.*

An early colonial writer named John Lawson claimed that Indian children in North Carolina were often cured of a craving to eat dirt by feeding them roast bat on a skewer (though Lawson doesn't say whether feeding dirt to adults might possibly cure their urge to roast a bat and make a child eat it). More recently, McGuane's raucous comedic novel *The Bushwhacked Piano* describes the scheme of a certain C. J. Clovis, former fat man and one-legged con artist, to contract with mosquito-plagued municipalities for the erection of bat towers—each fully stocked with 1,500 bats dyed Day-Glo orange—for solving the local bug problem. The bats would devour the mosquitoes, theoretically, while the cheerful dye allowed citizens to watch their investment in action. And the fictional C. J. Clovis (as McGuane himself probably knew) is but an echo of historical actuality: During the 1920s, a Dr. Charles Campbell, of San Antonio, proposed "municipal bat roosts" to control mosquitoes and thereby reduce malaria in the American South. At least one of those Campbell-style roosts was built, on Sugarloaf Key in Florida, by Mr. Richter C. Perky, who ran a fishing camp. Although accounts differ as to whether Perky ever actually stocked his tower with imported bats or merely baited it to attract the native ones, orange Day-Glo dye can safely be ruled out. But wait. Even this isn't all.

Practical bats have just hit the news again. A recent report in *American Heritage* magazine reveals that the United States government, during World War II, had a plan to use Mexican free-tailed bats for firebombing Japan.

The idea was to refrigerate these bats into hibernation, see, fit each with a small payload of napalm and a little-bitty parachute, see, drop thousands like that from planes over Japanese cities, see, and hope for the best. No I'm not making this up. Your government. The research cost $2 million.

Clearly the bat has captured human imaginations, and that may be because it seems triply oxymoronic: a flying mammal that sees in the dark by listening to its own silent screams. It is in truth an extraordinary animal, equipped with some startlingly sophisticated evolutionary adaptations and represented around the world by a wide variety of different forms. If the bat is grotesquely improbable, so is Pablo Picasso.

Chiroptera is the collective name for this order of mammals, and the main defining characters are familiar: wings formed by thin flaps of skin stretched between hugely elongated fingers; the habit of feeding by night and resting by day; hind feet that lock closed automatically when suspending the body's weight upside down; and a system of echolocation (at least among the Microchiroptera, one of two suborders), whereby the bat uses varying echoes from its own ultrasonic calls to find prey and steer itself through darkness. Anyone who has ever sat on a suburban porch while summer twilight faded has seen evidence of that last feature's uncanny efficiency. The erratic diving and swerving and swooping of those small black shapes reflect feeding success, not confusion. Their brains process the echo data at rates unimaginable to us—though some researchers on the human brain have been eagerly studying how they do it—so that a cruising and squeaking bat, while avoiding an array of tricky obstacles, can catch one third its own weight in flying insects within half an hour. Furthermore, they may have developed this sonar as much as 50 million years before we reinvented it.

The female bat possesses a single set of mammaries, from which her young are tenderly nursed, and that fact among others led Carl Linnaeus (the great Swedish classifier) to suppose they were very closely related to humankind. Actually, bats seem to have evolved from some small earthbound insectivorous mammal, a common ancestor linking them with moles and shrews. We don't know for sure, since the earliest bat fossil (from Wyoming) is fully batlike, and no trace has been found of a transitional form.

But if the assumption about cousinhood with the insectivores is correct, it only highlights still more the uniqueness of bats, because they have gone—in complexity, in diversity, in longevity—so far beyond their relatives. Moles and shrews still feed almost exclusively on insects, while various bat species (especially among the Megachiroptera, that other suborder) have attained much larger sizes and diverged into diets of fruit, nectar and pollen, fish, other bats, small birds and rodents, lizards, and blood. Moles and shrews have remained restricted to specialized environments, while bats have dispersed across every tropical and temperate area of the planet. One genus of bats, comprising sixty species, is more widely distributed than any other genus of mammals except the genus *Homo*. Most striking, though, is the matter of age. A shrew in the wild can expect a life span of one year, under ideal conditions maybe two. Bats commonly live ten years and longer, in some cases twenty, and achieve this longevity thanks to large measures of sleep and hibernation. One informed guess is that a bat might spend five sixths of its total life just hanging there, sound asleep.

Which seems fairly inoffensive. The bad reputation results, at least in part, from a small group of South American bats classified as the family Desmodontidae—the vampires. Contrary to popular notion, these creatures do not grow as large as ravens, do not possess hollow fangs for sucking, do not usually victimize humans, and do not inhabit Transylvania, or any other part of Europe. Their way of life entails sneak attacks on Brazilian cattle, delicately nipping open a vein with sharp incisors and then lapping away at the flow of blood with a dainty tongue. Also, rather oddly, they prefer to land a discreet distance from the intended cow and make their final approach on the ground, back hunched up high, tiptoeing along like some big-eared tarantula wearing a guilty smirk.

There, I concede, is an unsavory sort of bat—though perhaps even the Desmodontidae deserve credit for a certain roguish charm. Anyway, nobody ever suggested training vampires to serve as official United States weapons of war. That distinction was reserved for *Tadarida brasiliensis*, the Mexican free-tailed bat.

Tadarida brasiliensis offered one major advantage as tactical weaponry over other potentially deployable bats: abundance. There were 100 million of them roosting peaceably in just a few Texas caves.

Not even Tom McGuane and Albertus Magnus and Richter C. Perky all brainstorming together with Jack Daniels and George Dickel could have dreamed up an idea so robustly demented as this napalm-bat thing. It took a dental surgeon from Pennsylvania named Lytle S. Adams. Seems that Dr. Adams was driving home from a vacation in New Mexico, where he had gazed wide-eyed at millions of *T. brasiliensis*, like one continuous pelt of lumpy brown fur, covering for acres the ceiling of the Carlsbad Caverns, when news of Pearl Harbor reached him. In first froth of patriotic outrage and desirous of doing his bit, Adams thought of those bats. In less than two months, as the *American Heritage* article has it, Adams "somehow got the ear of President Franklin Roosevelt and convinced him that the idea warranted investigation." Under the circumstances, "somehow" seems rather tantalizingly elliptical, but maybe FDR needed a little dental surgery and Dr. Adams pitched his idea before the gas had entirely worn off. Next he managed to interest an eminent Harvard chiroptologist (a bat expert, not a foot doctor) named Donald R. Griffin, and before long the National Defense Research Committee had signed on as a sponsor. By now it was known as the Adams Plan. Eventually the army's Chemical Warfare Service, the NDRC, and the navy (no reason submarines couldn't release bats too) were all implicated in the buffoonery.

The first field tests were held at a remote airport in California on May 15, 1943. These were also, apparently, the last field tests. Adams and his colleagues discovered that *T. brasiliensis* could not always be put into hibernation, nor brought out of it, as promptly as might be convenient. And that the parachutes were a little too bitty. And that the incendiary capsules were a little too large. Groggy bats were tossed out of a plane. Many broke their wings. Some hit the ground without waking at all. It was a waste of innocent animals.

Yet there was poetic justice. A few other bats, armed on the ground with live napalm units but spared the lethal jump, escaped from their handlers. These escapees flew off toward the nearest buildings—as indeed they were supposed to do, though preferably in Japan—which happened to be the airport hangars. The hangars thereupon burned. So did a general's automobile.

It did not seem auspicious to NDRC officials. The Adams Plan, in mercy to bats and chiroptophiles everywhere, was canceled. And we can guess that Shakespeare himself would have appreciated the shapeliness of that denouement: *Fair is foul,* said the three witches, *and foul is fair.*

PROPHETS
AND PARIAHS

The Excavation of Jack Horner

"I DON'T GIVE A SHIT *what* killed the dinosaurs," says John R. Horner. Strange talk for an eminent paleontologist, but not out of character for this particular one. He is exaggerating his natural brusqueness only a little, in the interest of stressing a point. "They dominated the earth for 140 million years. Let's stop asking why they *failed* and try to figure out why they *succeeded* so well." From Horner's perspective, the entire Mesozoic era—during which the dinosaurs appeared, flourished, diversified, rose to supremacy among all terrestrial creatures, and then, somewhat abruptly, disappeared—is a Horatio Alger story, not a murder mystery.

Jack Horner's perspective is unconventional but authoritative. His recent fossil discoveries, and the surprising deductions toward which those fossils have led him, are being followed raptly by paleontologists all over the world. With his scruffy beard, longish hair, balding pate, he looks like a skinnier and younger version of the actor Warren Oates. On location, let's say, for a good-humored film about rowdy and disreputable prospectors. But in fact Horner is one of a trio of men—John Ostrom and Robert Bakker are the others—who during the past fifteen years have been drastically reshaping our understanding of dinosaurs. Ostrom is a venerable professor at Yale. Bakker has lately gone from Harvard to Johns Hopkins. Meanwhile Jack Horner sits, wearing a plaid flannel shirt and beaten-down running shoes, in the basement of a small museum in a place called Bozeman, Montana.

Like Richard Leakey, with his study of early mankind in northern Kenya, Horner has stepped suddenly into the front rank of scientists in his field despite a near-total lack of academic credentials. He never bothered to finish college. Never went to grad school. Doesn't read German or Russian. Knows almost nothing about computers. Unlike Leakey, Horner did not even have the advantage of famous scientist parents; his family owned a gravel-and-concrete business in Shelby, Montana. Horner is simply a brilliant and dogged bone-hunter, a field man, a natural, with a keen brain for imagining the ecological particulars of an age 70 million years gone. He has a nose for fossils and a head full of provocative ideas.

On a bare hillside not far from the Teton River in northwestern Montana, Horner and his field associates have unearthed a nest, roughly six feet across, containing the bones of eleven baby dinosaurs. In the same vicinity they have also found other nests, several more babies, and the fossilized remains of more than three hundred dinosaur eggs. Throughout the whole history of fossil collection, dinosaur eggs and juveniles have remained breathtakingly rare; no other nest full of hatchlings has ever been found. Consequently, there has been a tantalizing absence of just that sort of evidence necessary to answer certain crucial questions—questions about dinosaurian breeding habits, patterns and rates of growth, behavior among others of their kind. Jack Horner now has that sort of evidence.

Based on his finds, Horner believes that at least one group of dinosaurs were sociable, relatively intelligent, warm-blooded, and solicitous toward their own infant offspring. It's a little like announcing five centuries ago that the Earth isn't flat after all.

Warm-bloodedness, nesting in colonies, and extended parental care are all generally nonreptilian attributes, associated rather with mammals and birds. Reptiles are cold-blooded. They don't (except in rare and disputable cases) tend their young. They don't show advanced social behavior. Reptiles as we know them just don't act in the manner that Horner's nest-field seems to indicate.

But maybe the dinosaurs were not nearly so reptilian as tradition, and eight generations of paleontologists, have decreed. Maybe, suggests Jack Horner, they were something utterly different.

In more senses than one, Jack Horner grew up among dinosaurs.

The wild country of Montana has always been a bone-digger's paradise, partly because its hillsides and gulches have remained almost undisturbed by human development, more basically because this happened to be a place where great numbers of dinosaurs lived and died and where the sedimentary deposits in which their bones became fossilized have latterly been lifted and cracked open near the landscape's surface. Toward the end of the Cretaceous period, 70 million years ago, what is now the Midwest and the Great Plains was covered with a vast inland seaway, with central Montana elevated slightly along its western seacoast. Dinosaurs thrived in that swampy coastal zone, and when an individual died, sediments washing down from the newly burgeoning Rocky Mountains were liable to bury it. Finally the seaway withdrew, and the Cretaceous sediments were overlain with more recent strata; as subsequent epochs passed, erosion cut down through those strata, crustal pressures buckled and tilted the land, and in many places the Cretaceous deposits were reexposed to daylight. The result is a rich hunting ground for fossils, an enormous boneyard dating from exactly that time at which the dinosaurs hit their peak.

Back in 1855, the first dinosaur fossils to be found and described in the western hemisphere were taken from beds along the Judith River, not far from Fort Benton, Montana. In 1902, the modern world's first glimpse of *Tyrannosaurus rex* came from a dig near Jordan, south of Fort Peck. Jack Horner spent his boyhood at large in this terrain. He found his own first dinosaur bone when he was eight. A systematic kid, he used white paint to label the fist-sized chunk as specimen "104-A" among a boy's box of fossils.

"Did you save that bone?"

"Yup," Horner says.

"Do you still have it?"

"Yup."

"What is it? What part of what sort of animal?"

"I don't have the slightest idea."

Horner struggled through Shelby's only high school, and it would be understatement to say that in the classroom he showed no promise of future scientific renown. Languages, for some reason, were especially a problem. "Took me two years to manage a D in Latin One." Nevertheless he went on to the university, down at Missoula, hoping to do a degree in geology. His father harbored a dream, on Jack's behalf, of the career of a mining engineer. Jack himself was still dreaming about fossils. More specifically, about dinosaur fossils.

"Dinosaurs are really neat animals," he says even now, shamelessly ingenuous in his enthusiasm. "I mean, dinosaurs are *really neat* animals." Often enough he discusses them in the present tense, hypothesizing details about certain species or families as would any wildlife biologist: "A baby hadrosaur has very little to protect it."

But his initial try at the university ended sourly. "I'm a product of the sixties," Horner says with a glint of perverse pride—and his personal details support that self-analysis, since there's no better way to qualify as a true child of those times than by what befell him next. He flunked out of college in 1965 and was immediately drafted by the marines.

"Everybody thought that the Marines didn't draft. Remember? That's what I thought too. *The marines?*"

They sent him through something called "para-frog" training in Okinawa, where Horner was taught to leap out of airplanes over water, wearing a parachute on one part of his body and scuba tanks on another. Characteristically upbeat about personal matters, he counts himself lucky: Despite the training, he was never ordered to jump into an ocean during combat. Instead he jumped into jungle. Most of his thirteen months in Vietnam were spent on "force recon" duty. He would be dropped into the DMZ or some other feverish corner of Vietnamese jungle, with a small team or alone, carrying minimal firepower but a strong radio, and simply stay out there, discreetly, avoiding combat but reporting back south about whomever and whatever he saw.

During one of these solitary patrols, near Quang Tri, just south of the DMZ, he encountered a pair of North Vietnamese students. They were taking instruction at a Buddhist temple. Walking out of the jungle, Horner had seen the temple and was curious. He set his rifle down at the front door, because that seemed the courteous thing, and entered. Several Buddhist monks were there, with this pair of students; the monks were teaching them English and a smattering of biological sciences. The two students, Horner recalls, were the first people in all of Vietnam—Vietnamese or American—with whom he could talk. "*Really* talk. About more than hat size," says Horner. "Or what an M-16 could do to the human body. 'Yew ever see what a M-16 kin do to the human body?' That was always a favorite. So these two students, well, we just started talking. They were, literally, the most intelligent people I met in Vietnam." He told them a little about himself. Told them he was from Shelby, Montana. One of the North Vietnamese students said, "Is that close to Butte?"

The best Vietnam duty of all, to Horner's taste, was when he was left by himself to spend a week or two on "recon station," manning some unprotected little lookout post in the midst of some ungodly forward zone. "I liked being alone in the jungle," he explains. Surrounded by wild animals and crazy vegetation. A course of nature study. Not so different, he claims, from being home in the outback of Montana. Then one day he called in an artillery barrage but gave the wrong coordinates and collected a leg full of shrapnel when the American cannons shelled him instead of the enemy.

After Vietnam, Horner went back to the University of Montana, floundering as hopelessly as before. He was still fascinated by geology and paleontology, but for a degree in those subjects he was required also to pass courses in math, liberal arts, and a couple of foreign languages. The language requirements in particular were daunting. "I was in Russian class for three days before I figured out it was second term." At one juncture, Horner recalls, his grade-point average was so low it could be rounded off to zero. Out again, in again, out again, yet during all these years of frustrating academic travail, Horner was still going back up each summer, or whenever possible, to the Cretaceous formations in central Montana. He was digging and collecting with a fanaticism derived from sheer enjoyment. His determination, his love for being outdoors on the Montana landscape, his gift for reading rock, his stamina for crawling around in coulees on bruised hands and knees for hours at a time with finely focused attention—all these were making him a highly experienced field paleontologist, whatever the college records might say.

In 1973 he left the university altogether and began driving a gravel truck. Stone is a leitmotif throughout Horner's life.

Not long thereafter he moved up to an eighteen-wheel tractor-trailer rig, hauling tanks full of liquid fertilizer all over the state. He was paid by the day, but there was one incentive for making good time. "I always kept an eye open for Cretaceous rock. When I'd come to what seemed like a fossilly area, I'd just stop, unhook my trailer, and drive off across the badlands in that tractor. To look for dinosaur bones." Yet in Horner's mind the truck driving, even on these terms, was never more than an interim situation. During the same period he was mailing job-query letters to every paleontological museum in the country.

In 1975 he was hired by Princeton University to work as a fossil preparator (the paleontological equivalent of a dental hygienist), cleaning and gluing specimens that had been found by other people and were to be studied by other people. Faculty scientists, folk with Ph.D.'s. Horner was abundantly overqualified as a preparator, having done the same sort of chores in support of his own private studies for most of the past two decades. Nevertheless he stayed at Princeton for seven years, polishing his skills, learning the ways of museum work, earning a little autonomy, expanding his role by increments, and getting up and down the East Coast for a close look at every important dinosaur collection from Harvard to the Smithsonian. He also spent his vacations each summer out in Montana, gathering more fossils from the gulches and bluffs he knew well and thereby greatly enriching the Princeton collection.

One other significant matter was unearthed during those Princeton years, not a fossil but a fact. Thanks to a campus poster and then an exam, Horner learned for the first time that he suffered—and always had—from dyslexia. That cast some light on the inaptitude for languages, the academic struggles, the strong preference for fieldwork. Words on a page shifted and twisted and tangled themselves before Jack Horner's eyes. But a bone was a thing of solidity and eloquence.

In 1978, still under the Princeton aegis, he went back to north-western Montana, back to the same geologic formation where he had found 104-A, back to a bone-hunting partner named Bob Makela whom he had known since the time in Missoula, and together these two aging hippies made a world-class paleontological discovery.

Throughout human history until the late eighteenth century, mankind had no inkling that any such beasts as the dinosaurs had ever existed. Fossil skulls had turned up in a few places—Scythian gold mines of the seventh century B.C., for instance—but had either been ignored, left unexplained, or ascribed to mythical beasts such as the griffin. Only in 1841 did an Englishman coin the word "Dinosauria," lumping certain weird, newfound fossils into a category whose name translates as "terrible lizards." Actually they had been neither lizards (a distinct group of reptiles, unlike either dinosaurs or crocodilians) nor, most of them, very terrible. Many were large herbivores, pacific creatures making their livings in roughly the same way as a modern moose, or an elephant, or a giraffe. Even *Tyrannosaurus rex* may have been less the ferocious and implacable predator, as commonly portrayed, than a lazy and opportunistic omnivore, feeding on carrion or weakened animals or whatever was most convenient, as a grizzly bear does today. But for another 120 years, the conventional view of dinosaurs, both in popular presentations (like Disney's *Fantasia*) and among scientists, remained unshaken. According to that view, the meat-eating species were fierce predators that walked erect on hind legs; the vegetarians were huge gawkish dolts, slow-moving and vulnerable; and all of them were simply magnified variations on the anatomy and physiology of a lizard. Cold-blooded. Mentally dim. Lacking any hint of advanced social behavior.

Finally a few scientists rebelled. That traditional view was not only unsupported by fossil evidence, they said; it was downright self-contradictory.

In 1969, John Ostrom told a conference of paleontologists: "The evidence indicates that erect posture and locomotion probably are not possible without high metabolism and high uniform temperature." About the same time Armand de Ricqlès, a bone specialist in Paris, noticed that the internal structure of many dinosaur bones seemed to resemble mammal bones more closely than lizard bones. During the next several years Robert Bakker assembled a fuller framework of evidence that pointed the same way and published a pair of revolutionary papers in the journal *Nature*. According to Bakker, the dinosaurs had been warm-blooded. Some of them, to help maintain their thermal stability, had even developed an insulating layer of feathers. In their physiology, and most likely too in their behavior, they were advanced far beyond any lizard on Earth today. In fact, argued Bakker, they should not even be included among the reptiles. These animals had evolved into something distinct. Furthermore, wrote Bakker, "the dinosaurs never died out completely. One group still lives. We call them birds."

Following this line of thought, what Jack Horner and Bob Makela found on that Montana hillside was a great teeming dinosaur rookery. They had the first evidence of extended parental care, nesting in colonies, and elaborate social behavior (three attributes linking dinosaurs with birds) that was ever uncovered to human view.

Horner calls the site Egg Mountain, in a spirit of ironic but grateful homage. Actually it is only a gentle knoll, one among many out in this rolling terrain of sparse scrubby grass and hillocks and coulees cutting down into a fossil-rich layer of sedimentary rock known as the Willow Creek Anticline. The real mountains loom up in the west, a towering wall of dark peaks and cliff faces not more than a dozen miles off, snow-covered nine months of each year. That stretch of mountains, called the Sawtooth Range, is the easternmost front of the Rockies along this northern part of their length, the very juncture line where the great midland prairies come to a sudden halt, running smack up against the roofbeam of the continent. A few miles up the gravel road from Horner's Egg Mountain is another anomaly, Pine Butte Swamp, now protected by the Nature Conservancy because of its ecological uniqueness, a northern fenland of wolf willow and bog bean tucked flush against the base of the Rocky Mountain Front. The Pine Butte area is interesting to a biologist for forty reasons but noteworthy to any layman for one thing: It is the only place in the lower forty-eight states where *Ursus arctos*, once the most formidable beast on the American landscape, still ventures down onto prairie.

"This is the last place in America," says Jack Horner, "that has the grizzly bear still in its original habitat. Out on the plains. Do you know what it's like to be on your knees, looking for dinosaur bones—and at the same time you have to look over your shoulder, watching for grizzly?" His face contorts to a lopsided smile. "It's exciting." Bears wander over occasionally from Pine Butte to forage for roots or hunt rodents on the hillsides around Egg Mountain. "You come across a paw print, a *fresh* print, like eight inches long. And that land out there is just *open*. Nowhere to go. Not a tree to climb for miles." Another large grin.

Horner was on his knees like that, watching for small bones in the dirt and for big furry shapes over his shoulder, when he and Bob Makela made their historic discovery. On the side of Egg Mountain, in a bowl-shaped depression of brown mudstone, they found the skeletons of eleven baby dinosaurs of the Hadrosauridae family. The hadrosaurs were a group of semiaquatic herbivores, also called "duck-billed" dinosaurs for the slightly comical shape of their plant-gathering jaws, and though adult hadrosaurs were well known from Montana and elsewhere, neither complete juvenile specimens nor eggs had ever been found. Close by the first eleven were another four skeletons of the same type and size.

The depression was unmistakably a nest. Patterns of deep wear on the teeth showed that these babies had been feeding, and for a longish period—yet here they were, in a crowded jumble, still clinging to the cradle. They seemed to have died from neglect; suddenly orphaned, perhaps, at an age when they weren't yet capable of going out to shift for themselves. In a paper published in *Nature*, Horner and Makela wrote: "The fact that 15 baby hadrosaurs had been feeding, and had stayed together for a period of time, indicates that some form of parental care was administered for, if the young were confined to the nest, food must have been brought to them." If so, those young hadrosaurs and their doting parents were unlike any reptiles known in the world today.

Horner and Makela described the new species and named it

Maiasaura peeblesorum. The *peeblesorum* was in thanks to a family named Peebles, ranchers on whose land the find had been made. *Maiasaura*, according to Horner, means "good-mother reptile."

The excavations on Egg Mountain and in the surrounding area have continued for seven years, with no sign yet that this rich vein of fossils is even beginning to play out. More nests have turned up, more juveniles, and at least three hundred whole or partial eggs. Adults of *Maiasaura peeblesorum* have been found, as well as portions of adults from two other dinosaur species, one of which seems to have been a smaller carnivore, a swift creature that may have preyed upon young *Maiasaura*, snatching babies out of the nest when there was a lapse of parental protection. And along a certain ridge above Egg Mountain, stretching for more than a mile, is what appears to be a continuous, staggeringly abundant deposit of hadrosaur bones. Three thousand pieces have already been taken from one little trench; by extrapolation, the entire ridge deposit might contain several million. That sheer volume of contemporaneous fossils suggests that a vast herd of hadrosaurs, hundreds of animals, once gathered here sociably in a huge clamorous breeding colony, a rookery, finding security in numbers for themselves and their nestlings, in much the way penguins do today.

In 1982, Horner left Princeton. He accepted a position as curator of paleontology at the Museum of the Rockies, a modest institution connected with Montana State University, in Bozeman. The salary is meager. The library resources in paleontology at MSU are meager. The walls have no ivy. It's a museum where hadrosaur specimens share their end of the basement with a dry old Conestoga wagon. All of which is fine with Horner, who simply wanted to get back to Montana. The editors of *Nature*, in London, will not worry about his return address.

Through the winters—and out here they are long ones—he now studies his specimens, writes papers, teaches. Then in early June he moves north, with his teepee, to a campsite near Egg Mountain. In company with his old friend Makela (a science teacher at the only high school in Rudyard, Montana) and a few dozen assisting volunteers, he digs and scratches at the ground. The camp's crucial field-season supplies include a rented jackhammer, short-handle picks, ice awls, delicate brushes, and 150 cases of beer. For three months Horner is at large in the wild among *Maiasaura peeblesorum*, *Tyrannosaurus rex*, and the grizzly.

I asked Jack Horner if he could imagine any situation in life that he might prefer to the one he occupies now. He thought for a moment, carefully, and then said: "No."

The Lives of Eugène Marais

A JELLYFISH IS SOMETHING much more than the sum of its parts, but that hasn't always been so. Early jellyfish ancestors followed simpler arithmetic. They were precisely, and only, the sum total of a grouping of similar cells; they were in fact colonies of individual unicellular animals, each individual not terribly different from an amoeba. In the primordial gumbo of Precambrian oceans, the consensus for togetherness came about first, then division of labor, finally morphological specializations that fitted certain of the individuals for certain tasks. Some members of the colony went into service as gut lining, some as tentacles. This pattern of evolution toward multicellularity, not uncommon in the history of life, has been labeled *amalgamation*. The principle is as familiar as the print on a dime: *e pluribus unum*. Many simple lives fused into one complex life. Sponges evolved the same way. So did sea anemones and hydras and others of that gooey ilk. And so also, if we are to believe a charming crank named Eugène Marais, did an animal known as the termitary.

A termitary is a colony of termites.

In South Africa, where Eugène Marais spent most of his years, the predominant sort of termitary consists of sand particles heaped up like a giant pointy anthill, glandular secretions applied as mortar, fungus gardens kept damp in hidden compartments, and the moving bodies of uncountable individual termites. There is also one termite queen, swollen grotesquely with ovulation, too fat to move, ensconced and well tended within a royal chamber. Such a termitary might be forty feet high and hundreds of years old; it might include more than a million termites. A termitary found in the Limpopo Valley, according to careful measurements made by an engineer friend of Marais's, reportedly incorporated 11,750 tons of earth. And that mountainous pile of slobber-glued sand, with its intricate system of passages and rooms and ventilation ducts, with its hothouse mushroom patches, with its million living constituents clambering everywhere, was in reality—so Marais argued—a single animal. He was quite serious.

Eugène Marais was born in 1872 near Pretoria, and within the space of sixty-four years he lived more lives than a Hindu cow. He was at various times a naturalist, a newspaper publisher, a lawyer, a journalist, a medical student, a smuggler of munitions, and one of the first important vernacular poets in the Afrikaans language. He was also a morphine addict and a suicide. Beyond these details, given vaguely and sometimes contradictorily in a very few sources, the facts of his life are little known. If Eugène Marais hadn't existed, it would have been Jorge Luis Borges who invented him. But he did exist. That much we know for sure, because he left behind a matched pair of posthumously published books too concrete and too bizarre to have been imagined in fiction.

These are *The Soul of the Ape* and *The Soul of the White Ant*. It is entirely typical of the warps and wobbles of factuality throughout the Eugène Marais story that the first of the two is not about apes, the second is not about ants, and the pair were written in two different languages.

At age nineteen, Marais was the editor of a Pretoria newspaper called *Land en Volk*, and two years later he owned it. Pretoria in those days, just before the Boer War, was the capital of the Transvaal Republic and the site of its parliament, the Volksraad. Reporting and editorializing on parliamentary tomfoolery for his paper, Marais evidently was so scathing that, in the words of his son, "he had the distinction of being expressly excluded from the press gallery by a resolution of the Volksraad." Soon after that he stood up against Paul Kruger, the country's dictatorial president, who was taking steps to repress public gatherings and the press, in mind of turning Transvaal into an equatorial Prussia. Suddenly, for his meddling, Eugène Marais went to trial accused of high treason. Before the Supreme Court at Pretoria, he beat that charge.

During this early period as a political journalist, Marais was already developing the other interests that would later make him a fanatically observant naturalist, notable for his arcane sympathies. While running the newspaper, according to his son, Marais showed a strong affinity for animals and "was never without tame apes, snakes, scorpions, and the like." One of his favorite pet scorpions, by account of Marais himself, was a formidable female almost six inches long. This creature once attacked and killed an adult chicken in ten minutes. But she would sit on Marais's hand, grip him kittenishly with her claws, hold back her sting. Marais wrote: "She liked being scratched gently."

In 1894 he married a young woman who bore the one son and then died. About this time, taking the loss very hard, Marais began his morphine habit. And the next year—in the first of his abrupt metamorphoses—he went to London, with the idea of studying law or medicine, or both.

After four years of medical training—and again this is typical—he somehow became a lawyer. Then the Boer War broke out. As an enemy national during wartime, he could remain in London only on parole status. He left. When the war ended in 1902, with the defeat of his people by the British, Marais was back in Africa, preparing to smuggle a load of explosives and medical supplies across the Limpopo River to embattled Boer forces. About then he was hit with a bad attack of malaria. The supplies were buried, and he limped back to Pretoria. There he settled in for a phase of quietly practicing law.

Now the second metamorphosis: While living the life of a local attorney, he emerged almost magically as one of South Africa's most influential poets. The Afrikaans language was a slangy variant of Dutch, still fresh and unrecognized in those days, and Eugène Marais with his lyric poems seems to have done for it something of what (in a much grander way) Dante did for Italian. He showed that it was supple enough, beautiful enough, to hold art. His poem "Winter Nag" has been called the heraldic beginning of the new Afrikaans movement in literature. Today in South Africa he is chiefly known, despite the two visionary books on animal behavior, as an Afrikaans poet.

Naturally, after a few years of lawyering he was bored and disgusted. Metamorphosis number three: He retired to a remote gorge in the mountainous Waterberg district, built himself a hut, and lived there for three years in the company of a large troop of chacma baboons.

Long afterward he described that baboony time in a letter: "I followed them on their daily excursions; slept among them; fed them night and morning on mealies; learned to know each one individually; taught them to trust and to love me—and also to hate me so vehemently that my life was several times in danger. So uncertain was their affection that I had always to go armed, with a Mauser automatic under the left armpit like the American gangster! But I learned the innermost secrets of their lives." Those behavioral observations, and the innermost baboon secrets Marais felt he had deduced, became the basis for a book which he hoped would be his masterpiece, but which he never finished. A partial manuscript was finally published as *The Soul of the Ape*—though not until 1969, and then only through the help of the late Robert Ardrey, whose own *African Genesis* had been dedicated to Marais.

Marais harbored a high-flown opinion of the ideas in his baboon manuscript: "I have an entirely new explanation of the so-called subconscious mind and the reason for its survival in man. I think I can prove that Freud's entire conception is based on a fabric of fallacy." The kernel of his argument is that the human unconscious, as discovered and described by Freud, is nothing other than the older and more basic conscious mentality of prehuman primates, which has been pushed into the psychological background, but not eliminated, by the newly evolved human consciousness. In other words, the human unconscious is identical with—in Marais's choice of phrase, meant literally—the soul of the ape. It's a proposition that, to say the least, has few followers among modern psychologists.

The baboon watch ended when Marais's recurrent malaria and his morphine habit (possibly also loneliness) drove him back again to Pretoria. But throughout his three years among the baboons and perhaps (the tapestry of known fact is at this point especially threadbare) for seven years thereafter, Marais was, in addition to all else, a passionate student of termites. He spent long hours watching them. He performed experiments. He traveled to inspect unusual termitaries. He scratched his head and speculated. How, in a parched countryside, do millions of termites satisfy their constant need for water? Why do they grow fungus and gather hay? How does the queen get from one royal chamber to another, given that she's too large to fit through the door, incapable of moving herself, and seemingly too heavy to be lifted? For these mysteries and others, Marais found solutions—some right, some wrong but plausible, some cockamamie. Above all he posed and answered the question, What manner of thing is a termitary? It is an organism, he said; a single living animal.

The life of a termitary begins with the nuptial flight, when a winged male and a winged female—each dispersed from an existing termitary—meet and mate and dig a small nest for their offspring. This founding pair, from which all the millions of other individuals will be directly descended, are in Marais's view the "generative organs" of the termitary. The king remains small while the queen grows hugely distended and is before long laying 50,000 eggs every twenty-four hours. The offspring are mainly wingless and asexual, divided into worker and soldier castes; these workers and soldiers, according to Marais, constitute respectively the red and white corpuscles of the bloodstream. Deep within the termitary, fungus gardens serve as stomach and liver, wherein vegetable food from outside is left to be decomposed by fungi before the termites themselves eat it. And meanwhile the queen in her hardened chamber, giving off a pheromonic essence that inspires purpose and cohesiveness among all the population, is (besides being the ovary—Marais's theory is not without some wooliness) the brain. Kill the queen and, true enough, the entire termitary dies.

Marais concluded: "The termitary is a separate composite animal at a certain stage of development, and lack of automobility alone differentiates it from other such animals." It is, he added, "an example of the method in which composite and highly developed animals like the mammals came into being." That is, in the deep past, at some simpler phase of their evolutionary history, by amalgamation. This is his personal argument, remember, not mammalian or termite phylogeny as presented by more reliable authorities.

Marais's termite studies appeared, beginning in 1923, as a series of short articles in various Afrikaans newspapers and in a magazine called *Die Huisgenoot*. A final and definitive article was published by *Die Huisgenoot* in 1925. Written in Afrikaans, it would have been intelligible only to Boers, Dutchmen, and anyone who spoke a version of Dutch, such as Flemings.

Maurice Maeterlinck, an eminent European playwright and Nobel Prize winner, happened to be a Fleming. Evidently he saw the 1925 article. The following year Maeterlinck published a book titled *The Life of the White Ant*, expropriating the detailed observations of Eugène Marais, and his terminology, and his theory about termite amalgamation. The book was a success. Marais got no acknowledgment. He hadn't the money to press a lawsuit.

Downward spirals of morphine, self-pity, anger, depression. Nine years passed, and then a woman in London, Winifred de Kok, began translating Marais's own termite pieces into English for their eventual publication as *The Soul of the White Ant*. She corresponded with Marais, and the letters to her tell us most of what we know about his inner life. He seemed to take new hope. He seemed buoyant for the first time in years. He was finally to have vindication. In one letter he wrote: "You see that your kindly enthusiasm has infected me! . . . The thought of reaching a bigger public intrigues me."

Five months later he put a shotgun to his head and fired.

What are we to make of such a man, such a life, such a set of lives? Robert Ardrey has called him "the purest genius that the natural sciences have seen in this century." Well, no. The works left behind by Eugène Marais don't begin to support that claim. What, then?

As a newspaperman, he ruined himself by excessive and caustic candor.

As a lawyer, he was a good poet, or at least an influential one. As a poet, he was not Wallace Stevens. I once spent a day in the New York Public Library with several scarce old collections of his poems, and the stuff seemed to me pretty terrible. Maybe it was the translations. Maybe not.

As a descriptive naturalist, he was, at his best, wonderful. But even *The Soul of the White Ant* is a badly flawed book, with patches of metaphysical gobbledegook and lame guesses and non sequiturs mixed in among the wonders.

As a theorist of insect and primate evolution, he was, I think, more than a little brilliant and more than a little nuts.

So what's the bottom line? There is none. It's a reductive concept, *bottom line*, and Eugène Marais—less even than most humans—just can't be thus reduced. He does not lend himself to categorization, easy dismissal, unreserved adulation, or summary assessment of any sort.

Except maybe this. As an amalgamation of many individual lives, that polymorphic phenomenon to which adhered the name Eugène Marais was one exotically complex jellyfish. He was a man of parts. But he was something much more than the sum of them.

The Man with the Metal Nose

A GOOD FRIEND OF MINE CLAIMS, among other matters of wild and ornery personal ethic, that he chooses his friends strictly by smell. I'm not sure what this says about my personal odoriferousness, but I do understand, and endorse, his point in principle: The ineffable qualities are the ones that count, not the objective characteristics that can be captured in an introduction or on a résumé or during two hours' conversation over cocktails. Those qualities will answer the more crucial questions upon which friendship is based, such as *Would this person instinctively step between me and a charging elephant?* or *Can I trust him to borrow a book without turning the page corners down?* And the nose, being humanity's most under-developed sensory organ, is perhaps the only apt emblem for our groping and sniffling efforts to register the ineffable. Which is why I can't stop wondering about one particular nose that occupies an intriguingly prominent place in the history of scientific inquiry.

It was an artificial one, this nose, a prosthesis made of gold and silver alloy. It was worn by an aristocratic Danish astronomer of the sixteenth century, a portly and sybaritic man named Tycho Brahe, who had lost his own God-given schnozzle in a duel. History does not record whether the replacement was held in position by a leather thong (like that curiously similar one worn by Lee Marvin in the movie *Cat Ballou*), or if not, then how. We do know that all his life Tycho carried a small snuffbox full of ointment with which, like one of those people compulsive about Chapstick, he kept his metal nose constantly lubricated. There is likewise no testimony as to what purpose this cold piece of technology might have served. The surviving portraits of Tycho suggest that its role was not to support eyeglasses. Did it smell? Did it run? Could it be turned up disdainfully? Was it often out of joint? We'll never know. Tycho himself, with or without his peculiar nose, would probably be forgotten by history if it weren't for two important considerations. The second of these was a set of notebooks full of numbers, and we'll come to them in a moment. The first was a galactic event of large magnitude.

Step outside on a summer night and look off toward the northeastern part of the sky. Not far below the Little Dipper you'll see the constellation Cassiopeia, easily recognizable in the shape of a W. Back in early November of 1572, when Tycho Brahe was still a young amateur stargazer of twenty-five, a bright new star appeared suddenly in that constellation. Flaring into view, it shone with more brilliance than any other star, more brilliance than the planet Venus, so brightly that it could be seen even during daylight. Furthermore, it gleamed from a spot where, just a week earlier and throughout the centuries before, no star at all had ever been visible. This phenomenon posed a serious philosophical problem in the late sixteenth century, when Aristotelean cosmology as sanctioned by the Catholic Church decreed that the upper celestial spheres—everything out there beyond the moon—were absolutely immutable. On the fourth day of Genesis, God had created the lights in the firmament, and that was that. Now suddenly here was a big new dot of fire flaunting its power in Cassiopeia. The star attracted attention and concern, not just among astronomers and theologians. It was a popular event of mythic resonance. And it made the reputation of Tycho Brahe.

Tycho wasn't the first knowledgeable watcher to spot the new star, but he noticed it for himself one night before the news had gone public, and it left him agape. Over the next sixteen months, while the star changed color and rapidly dimmed, he performed a continuous sequence of very precise measurements, using a fine sextant he had crafted from walnut wood (the best available technology at the time, given that telescopes hadn't yet been invented). Those measurements allowed him to speak of this new star more authoritatively than anyone else in Europe. It was immobile relative to Cassiopeia, Tycho reported; it was not in the sublunary atmosphere but far beyond, amid the other stars; it was not a comet lacking a tail, as some thought, but a true star. Tycho's book, *De Nova Stella*, made him internationally famous. He had charted all apparent aspects of the star with surpassing accuracy; but he had no idea what the heck it was.

Today we know: a supernova explosion. Only five such events have been visible from Earth during the past thousand years, and of those, Tycho's in 1572 was the fourth. What it signified was that, some thousands of years earlier at a very great distance, a gigantic star (much larger than our sun) had come to the end of its life span—the hydrogen nuclei at its core all "burned" by fusion to form helium nuclei, and the helium further fused into still bigger nuclei. The star had then fallen into a terminal sequence of convulsions, alternately expanding and contracting, gravitational compaction seething down against rising internal pressures, which led to a final cataclysmic thermonuclear explosion. That explosion flashed out perhaps a billion times brighter than the normal intensity of the same star during its previous life. And the flash, having traveled across all the light-years of distance between, eventually reached retinas on Earth from the direction of Cassiopeia. Then, by 1574, it was gone. No one knew why. Not even Tycho Brahe.

But Tycho, having so faithfully measured and plotted the thing, was now a national hero in Denmark. The king gave him his own island, as well as lavish financial support with which to construct a great astronomical observatory that would be Tycho's private scientific demesne. Tycho built a castle in Gothic Renaissance style, with spires and gables and cornices, and at the apex an onion dome topped by a gilt vane in the shape of Pegasus. There were guest rooms and aviaries and fountains, formal gardens and neat orchards laid out within a perimeter wall, fish ponds, English mastiffs to stand guard, a paper mill, a print shop for his publications, and from ceiling to floor in the main workroom an oversized mural of Tycho himself. He called the place Uraniborg. The various chambers and towers he furnished with all the best astronomical instruments a king's money could order up: sextants of walnut, quadrants of brass and steel, armillary spheres ornamented with his own portrait, triquetrums and azimuth circles and astrolabiums—who knows what they all did. In this setting Tycho commanded his many assistants, threw grand parties for visiting nobility, rubbed ointment on his metal nose, and tossed scraps of meat to his attending dwarf, Jeppe, who served as official court fool. Tycho, in other words, was not a scientist in the ascetic vein.

But during the next twenty years at Uraniborg he also performed the most precise and potentially useful collection of continuous astronomical observations that mankind up to that date had achieved. Where other astronomers (including most recently Copernicus) had been casual and sporadic about their observations, Tycho was thorough, punctilious, indefatigable. Where others had tracked the planets with only their unaided eyes, or occasionally a primitive sextant, Tycho devised his ingenious new instruments. Where others watched for a few nights or a few months, then went inside to dream up more or less misguided theories, Tycho kept watch relentlessly for over two decades, all the while recording his careful notes. The large quarto volumes containing those notes were his treasure. His contribution to science lay in recognizing that serious astronomy required data-gathering of such precision and continuity, and in marshaling the financial resources, the elaborate equipment, the patience, to make it possible. But again, as with the star of 1572, Tycho never knew what he had.

He was not persuaded by the Copernican theory of celestial organization (which had been published quietly about fifty years earlier) and was dissatisfied with the old Ptolemaic view. So in 1588 Tycho announced his own version. Earth, according to him, was stationary in space, as Ptolemy had thought. The other planets, he said, moved in uniform circular motion around the sun. And the sun in turn orbited, pulling its satellites along, in a great graceful circle around Earth. This Tychonic system supposedly explained all the complex planetary motions that Tycho's sky-watching over the years had so accurately mapped. It was mathematically sweet and theologically acceptable. Its only drawback was that it was wrong.

After two decades at Uraniborg, where Tycho was a greedy and irresponsible landlord to the island's peasants, putting himself gradually into disfavor with the new Danish king, those munificent cash subsidies ended. So the astronomer felt obliged to pack up his gear and his entourage and leave. He went shopping across Europe for another royal patron willing to support him in similar high style, and two years later he found one: Rudolf II, king of Bohemia and Holy Roman Emperor. Tycho settled into a new castle just outside Prague, on the River Iser. Again there was money enough to pay for lordly living and a staff of assistants, among whom now was a twenty-nine-year-old German who had already earned modest recognition as an astronomer in his own right. This man's name was Johannes Kepler, and he had some ideas about celestial organization himself.

Kepler had abandoned everything to join Tycho in Prague for a single ulterior reason: He hungered to see the data in those precious notebooks. But Tycho let him go hungry, assigning Kepler to some demeaning lesser chores while refusing to share information with him as a colleague. Then, in October 1601, Tycho Brahe suddenly died. And Kepler got hold of the notebooks.

Within eight years, using Tycho's data, Johannes Kepler had formulated and published two laws that for the first time accurately explained the dynamics of our solar system, and thereby began the modern age in astronomy. The laws were as simple, once recognized, as they had been inscrutable before. First, said Kepler, the planets (including Earth) travel around the sun not in circles but in ellipses, great oval orbits with the sun nearer one end. Second, each planet moves not at uniform speed but at a velocity that changes according to its distance from the sun. Today those statements might seem unexceptional. But in 1609, how many minds could have guessed that God would design a universe using *ovals* and *irregular motion*?

Something more was at work here than just astronomical training, hard thinking, and Tycho Brahe's data. What else? In many of the great scientific discoveries there seems to have been an additional mode of perception that took up in the shadowy zone where pure rationality ended, a further faculty that helped point the way to a revolutionizing insight. The word "intuition" is sometimes applied but, like a paper label on a bottle, only obscures what's inside. Arthur Koestler, in his intriguing book on the early astronomers, calls it "sleep-walking." Einstein spoke in his own case of "the gift of fantasy." As a young man of twenty-three, Isaac Newton suddenly glimpsed his law of gravity in little more time than an apple would take to fall from a tree (though the literal falling-apple anecdote seems to have been apocryphal). Alfred Russel Wallace got the idea of evolution by natural selection (though that wasn't his term) with the same suddenness, during an attack of fever, after Charles Darwin had labored over the same question methodically for years. Watson and Crick found the structure of DNA using Tinkertoys, youthful cockiness, and someone else's x-ray crystallographs—crystallographs that until then had not been correctly interpreted. In each of these entries upon the ineffable, something more was at work than mere cerebration.

Likewise with Johannes Kepler. He shaped his inherited Tychonic data into a vision of cosmological order that was ingeniously simple, drastically unorthodox, and true. But Tycho himself, evidently, just didn't have the nose for it.

Animal Rights and Beyond

DO NONHUMAN ANIMALS HAVE RIGHTS? Should we humans feel morally bound to exercise consideration for the lives and well-being of individual members of other animal species? If so, how much consideration, and by what logic? Is it permissible to torture and kill? Is it permissible to kill cleanly, without prolonged pain? To abuse or exploit without killing? For a moment, don't think about whales or wolves or the California condor; don't think about the cat or the golden retriever with whom you share your house. Think about chickens. Think about laboratory monkeys and then think about laboratory rats and then think also about laboratory frogs. Think about scallops. Think about mosquitoes.

It's a tangled question that, in my view, isn't well suited to straight answers. Some people would disagree, judging the matter simply enough settled, one way or the other. *Of course they have rights. Of course they don't.* I say beware any such absolute certitude. Some folks would even—this late in the evolution of human sensibility—call it a frivolous question, a time-filling diversion for emotional hemophiliacs and cranks. *Women's rights, gay rights, now for Christ's sake they want ANIMAL rights.* Notwithstanding the ridicule, the strong biases on each side, it is a serious philosophical issue, important and tricky, with almost endless implications for the way we humans live and should live on this planet.

Philosophers of earlier ages, if they touched the subject at all, were likely to be dismissive. Thomas Aquinas declared emphatically that animals "are intended for man to make use of them, either by killing or in any other way whatever." Descartes held that animals are merely machines. As late as 1901, a moral logician named Joseph Rickaby, a Jesuit, declared: "Brute beasts, not having understanding and therefore not being persons, cannot have any rights. The conclusion is clear." But maybe no, not quite so clear. Recently, just during the past decade, professional academic philosophers have at last begun to address the matter more openmindedly.

Two thinkers in particular have been influential: an Australian named Peter Singer, an American named Tom Regan. In 1975, Singer published a book titled *Animal Liberation*, which stirred the debate among his colleagues and is still treated as a landmark. Eight years later Tom Regan published *The Case for Animal Rights*, a more thorough and ponderous opus that took position as a sort of companion piece to the Singer book. In between there came a number of other discussions of animal rights, including a collection of essays edited jointly by Singer and Regan. Despite the one-time collaboration, Peter Singer and Tom Regan represent two distinct schools of thought. They reach similar (not identical) conclusions about the obligations of humans to other animals, but the moral logic is very different, and possibly also the implications. Both men have produced some formidable work and both, to my simple mind, show some glaring limitations of vision.

Having spent the past week amid these books, Singer's and Regan's and the rest, I'm now more puzzled than ever. I keep thinking about monkeys and frogs and mosquitoes and—sorry, but I'm quite serious—carrots.

Peter Singer's view is grounded upon the work of Jeremy Bentham, the eighteenth-century British philosopher widely known as the founder of utilitarianism. "The greatest good for the greatest number" is the familiar, simplistic version of what, according to Bentham, should be achieved by an ethical ordering of society and personal behavior. A more precise summary is offered by Singer: "In other words, the interests of every being affected by an action are to be taken into account and given the same weight as the like interests of any other being." If this much is granted, the crucial next point is deciding what things constitute "interests" and who or what qualifies as a "being." Evidently Bentham did not have just humans in mind. Back in 1789, optimistically and perhaps presciently, he wrote: "The day *may* come when the rest of the animal creation may acquire those rights which never could have been withholden from them but by the hand of tyranny." Most philosophers of his day were inclined, as most in our day are still inclined, to extend moral coverage only to humans, on the grounds that only humans are rational and communicative. Jeremy Bentham took exception: "The question is not, Can they *reason?* nor, Can they *talk?* but Can they *suffer?*" On this crucial point, Peter Singer follows Bentham.

The capacity to suffer, says Singer, is what separates a being with legitimate interests from an entity without interests. A stone has no interests that must be respected, because it cannot suffer. A mouse can suffer; therefore it has interests and those interests must be weighed in the moral balance. Fine, that much seems clear enough. Certain people of sophistic or Skinnerian bent would argue that there is no proof a mouse can in fact suffer and that to believe so is merely an anthropomorphic assumption; but since each of us has no proof that *anyone* else actually suffers besides ourselves, we are willing, most of us, to grant the assumption. More problematic is that very large gray area between stones and mice.

Peter Singer states: "If a being suffers, there can be no moral justification for disregarding that suffering, or for refusing to count it equally with the like suffering of any other being. But the converse of this is also true. If a being is not capable of suffering, or of enjoyment, there is nothing to take into account." Where is the boundary? Where falls the line between creatures who suffer and those that are incapable? Singer's cold philosophic eye travels across the pageant of living species—chickens suffer, mice suffer, fish suffer, um, lobsters most likely suffer, *look alive, you creatures!*—and his damning gaze lands on the oyster.

The oyster, by Singer's best guess, doesn't suffer. Its nervous system lacks the requisite complexity. Therefore, while lobsters and crawfish and shrimp possess inviolable moral status, the oyster has none. It is a difficult judgment, Singer admits, by no means an infallible one, but "somewhere between a shrimp and an oyster seems as good a place to draw the line as any, and better than most."

Moral philosophy, no one denies, is an imperfect science.

Tom Regan takes exception with Singer on two important points. First, he disavows the utilitarian framework, with its logic that abuse or killing of animals by humans is wrong because it yields a net overall decrease in welfare among all beings who qualify for moral status. No, argues Regan, that logic is false and pernicious. The abuse or killing is wrong in its essence, however the balance comes out on overall welfare, because it violates the rights of those individual animals. Individual rights, in other words, take precedence over maximizing the common good. Second, in Regan's opinion, the capacity to suffer is not what marks the elect. Mere suffering is not sufficient. Instead he posits the concept of *inherent value*, a complex and magical quality possessed by some living creatures but not others.

A large portion of Regan's book is devoted to arguing toward this concept. He is more uncompromisingly protective of certain creatures—those with rights—than Singer, but he is also more selective; the hull of his ark is sturdier, but the gangplank is narrower. According to Regan, individual beings possess inherent value (and therefore inviolable rights) if they "are able to perceive and remember; if they have beliefs, desires, and preferences; if they are able to act intentionally in pursuit of their desires or goals; if they are sentient and have an emotional life; if they have a sense of the future, including a sense of their own future; if they have a psychological identity over time; and if they have an individual experiential welfare that is logically independent of their utility for, and the interests of, others." So Tom Regan is not handing rights around profligately to every cute little beast that crawls over his foot. In fact, we all probably know a few humans who, at least on a bad night, might have trouble meeting those standards. But how would

Regan himself apply them? Where does he see the line? Who qualifies for inherent value, and what doesn't?

Like Singer, Regan has thought this point through. Based on his grasp of biology and ethology, he is willing to grant rights to "mentally normal mammals of a year or more."

Also like Singer, he admits that the judgment is not infallible: "Because we are uncertain where the boundaries of consciousness lie, it is not unreasonable to advocate a policy that bespeaks moral caution." So chickens and frogs should be given the benefit of the doubt, as should all other animals that bear a certain degree of anatomical and physiological resemblance to us mentally normal mammals.

But Regan doesn't specify just *what* degree of resemblance.

The books by Singer and Regan leave me with two very separate reactions. The first combines admiration and gratitude. These men are applying the methods of systematic philosophy to an important and much-neglected question. Furthermore, they don't content themselves with just understanding and describing a pattern of gross injustice; they also argue, emphatically, that the injustice should stop. They are fighting a good fight. Peter Singer's book in particular has focused attention on the outrageous practices that are routine in American factory farms, in "psychological" experimentation, in research on the toxicity of cosmetics. Do you know how chickens are dealt with on large poultry operations? How veal is produced? How the udders of dairy cows are kept flowing? Do you know the sorts of ingenious but pointless torment that thousands of monkeys and millions of rats endure each year to fill the time and the dissertations of uninspired graduate students? If you don't, by all means read Singer's *Animal Liberation*.

My second reaction is negative. Peter Singer and Tom Regan, it seems to me, share a myopic complacence not too dissimilar to the brand they so forcefully condemn. Theirs is a righteous and vigorous complacence, not a passive and unreflective one. But still.

Singer inveighs against a sin he labels *speciesism*—that is, discrimination against certain creatures based solely on the species to which they belong. Regan uses a slightly less confused and clumsy phrase, *human chauvinism*, to indicate roughly the same thing. Both of them arrive, supposedly by sheer logic, at the position that vegetarianism is morally mandatory. To kill and eat a "higher" animal, they assert, represents an absolute violation of one being's rights; to kill and eat a plant violates nothing at all. Both Singer and Regan claim to disparage the notion—pervasive in Western philosophy since Protagoras—that "Man is the measure of all things." Both argue elaborately against anthropocentrism, while creating new moral frameworks that are also decidedly anthropocentric. Make no mistake: Man is still the measure, for Singer and Regan. The test for inherent value has changed only slightly. Instead of asking *Is the creature a human?* they ask *How similar to human is similar enough?*

Peter Singer explains that shrimp deserve brotherly treatment but oysters, so different from us, are morally inconsiderable. In Tom Regan's vocabulary, the redwood tree is an "inanimate natural object," sharing that category with clouds and rocks. But some simple minds would say: Life is life.

Alias Benowitz Shoe Repair

I FIRST HEARD ABOUT GEORGE OCHENSKI from a friend of mine who happens to be president of the Montana River-Snorkelers Association. We were in a fancy restaurant, as I recall, and there was wine involved. Ochenski had come to my friend's attention in the course of his (the friend's) presidential duties, which in strict point of fact are nonexistent. I should explain that the MRSA presidency is a purely honorary title, self-bestowed actually, because the MRSA is a mythical organization. This is quite different, please note, from labeling the organization itself nonexistent. Certainly the Montana River-Snorkelers Association does exist (mainly over wine and beer at various bars and restaurants, occasionally also around a campfire); it just isn't *real*. An actual mythical entity, then, the MRSA, of roughly the same ontological status as the NCAA national championship in football, or the domino theory of international relations. You should look into this fellow Ochenski, my friend told me. He can be reached care of Benowitz Shoe Repair, in a tiny town called Southern Cross, up in the Flint Mountains above Anaconda. Have some more cabernet, I said. But sure enough, it turned out to be true. Benowitz Shoe Repair is another mythical entity, existent in its own way but not real. George Ochenski is both mythical and real. Are you with me so far?

Ochenski must certainly be the preeminent river-snorkeler in the Rocky Mountains. He has talent, commitment, infectious enthusiasm, broad experience, state-of-the-art equipment, and a measure of lunatic daring. He has precious little competition. Most importantly, he has self-abnegating dedication to a larger purpose. Sometimes you have to snorkel a river, Ochenski believes, in order to save it.

So dedicated is George Ochenski, and so scornful of risk, that—if necessary to make a point—he is willing even to snorkel the Clark Fork River downstream from the Anaconda smelter.

A river-snorkeler, in case this isn't self-evident, is someone who swims downstream in a river with his face underwater, enjoying the ride, watching the scenery, breathing through a little tube. It's a lazy, hypnotic pastime best practiced on pellucid trout streams in midsummer. A few of us have been toying at it for years.

But George Ochenski does not toy. He jimmies himself into a full wetsuit, adds fins and a hood and neoprene gloves and a fanny pack holding three cans of beer, pulls a pair of skateboarding knee pads into place, defogs his mask, and jumps into rivers. Gentle rivers, and raging whitewater monsters. Last year, for instance, he did 38 miles of the Salmon River in Idaho without benefit of a boat. Also last year he leapt into the Quake Lake trench—an unusually steep and ragged stretch of the Madison River, created by rockfall during an earthquake, famous for biting kayaks in half—and nearly died. On that run his mask was ripped off six times while he tumbled head over teakettle through a garden of sharp boulders. The trench experience, George admits today, was a miscalculation. In Montana this kind of behavior does not pass unnoticed. By word, and more discreetly by the looks on their faces, people frequently tell him: *Son, you must be out of your everlovin' skull.* But they said that to Orville Wright, and they were wrong. Then again, they said it to Evel Knievel, and they were right. George Ochenski figures somewhere in between.

He has an enduring though ambivalent attraction to what he calls "death sports." Huge squinting grin from George as he acknowledges this ambivalence. Mountaineering. Ice climbing. Scuba. Never a major injury, never a bad accident—unless you count the time he fell 600 feet down a rock slope in the Alaska Range and did a self-arrest on his nose. Back in those years he traveled exotically for serious climbing, with generous sponsorship from the equipment companies, and took part in the first successful ascent of the west face of Alaska's Mount Hayes. Scaled some breathtaking frozen waterfalls. Around the same time, a consummate autodidact, he turned himself into an expert cobbler, because he wasn't satisfied with the professional repair work on his climbing boots; before long he was doing work for his friends too, and they had rechristened him, whimsically and metonymically, "Benowitz Shoe Repair." Today he mostly stays close to the little wood-heated cabin at Southern Cross, in the front of which stands a bass fiddle. The fiddle is a logical switch from tuba, which he played for thirteen years. Benowitz is a man of many skills.

Several years ago, in response to pressures both internal and external, he gave up the glorious climbing, thanked the sponsors, and settled down to being useful politically. He had come to feel that he owed something back to the mountains and rivers; meanwhile, there happened to be a certain crisis brewing near home. He now makes his living as an editorial assistant to an author of textbooks on environmental science. The cabin is filled ceiling-high with an eclectic library. On one wall is a quote from Congressman Ron Dellums, a statement of mixed metaphors and straightforward passion: "Democracy is not about being a damn spectator against the backdrop of tap-dancing politicians swinging in the winds of expediency." Ochenski himself, by disposition and habit, is certainly no spectator. Some people, particularly among interests on the opposing side, might still take him for a wild-haired, good-timing, reckless flake. They would be grievously mistaken. George Ochenski has an excellent brain, he has chutzpah, he has focus.

And in a small trailer up the hill behind his cabin, he has an Apple computer, its floppy disks full of damning information concerning the Anaconda Minerals Company.

On September 29, 1980, the Anaconda Company announced that it was closing its copper-smelting operations at the town of Anaconda. This came as a severe shock to the 1,000 smelter workers suddenly unemployed, and marked the end of a century of awesome environmental pillage. For one hundred years the Company (as it's known in Anaconda and Butte) had cut down forests, poisoned streams, smelted copper and other valuable metals, piled up vast mounds of slag, filled the air of the county with a sulfurous smog, and preserved its standing with the local community—despite such depredations—by dispensing regular paychecks. Now the economics of copper had shifted. Goodbye, thanks for everything. "The Company thought they could just lock the doors and walk away," says George Ochenski.

He and a few other Anaconda folk, some of them former smelter workers, think otherwise. They are after the Company like a fice dog after a bear. They have formed an enraged-citizens' organization, pressured the governor, pressured the congressional delegation, pressured the EPA. They want more than goodbyes. They want reclamation. They want accountability. At the very least they want precise information about the nature and magnitude of the poisonous mess left behind.

With sulfur dioxide no longer pouring from the smelter stack, the chief concern now is over toxic metals: lead, cadmium, mercury, zinc, copper itself, and especially arsenic. One hundred years of copper smelting have left various concentrations of some or all of these in the waters, in the plants, in the soil, in the animals of the county. George Ochenski and his compatriots want to know: how much? How much was dumped in the ponds, how much was buried, how much is still blowing free off the smelter site? How much is already in our lungs and our bones? How much is ingested with each rainbow trout from the Clark Fork River, if a person should be so lucky as to catch one of the surviving fish and so foolhardy as to eat it?

How much lead? How much cadmium? How much arsenic? The Anaconda Company no doubt devoutly wishes that these questions would go away.

Sometimes you have to snorkel a river in order to save it. Guided by this dictum, George Ochenski loaded his gear into the back of my car. It was late in the season, Labor Day weekend, with the air already growing cool. We paused briefly, where the gravel lane down from Southern Cross joins a paved county road, to check the

Benowitz Shoe Repair mailbox. Then George led me off on a pair of brief but illuminating tours.

We went to the Big Hole River, across the Continental Divide from Anaconda and clear of the war zone over heavy metals. The Big Hole is still a pellucid trout stream. We jimmied ourselves into wetsuits, added fins and hoods and neoprene gloves; I pulled George's one extra skateboarding pad into position over one of my knees, leaving the other to chance. Masks were defogged, snorkels adjusted, and we jumped in.

The view was beautiful. Trout and whitefish looked me in the eye, aghast, and skittered away. Sculpins darted discreetly for cover. I observed the differences in underwater behavior among three different species of stonefly. I gazed at the funnel webs of *Arctopsyche* caddisfly larvae, down between rocks in the fast water, which I had read about often but never before seen. I found a mayfly nymph equipped with an elephantine pair of tusks. We passed through a few modest rapids, where the current abruptly accelerated and the boulders came at us like blitzing linebackers who had to be straight-armed away. After two hours of cruising, we were nearly hypothermic, but the experience had been delightful.

Our other tour was to the Clark Fork River, downstream from the settling ponds into which the Anaconda Company has voided its years of industrial offal. "We're off to snorkel the Clark Fork," George told a friend as we pulled out of town. The friend looked puzzled. Huge squinting grin from George. "Then we'll come back and glow in the dark."

We snorkeled a long section of the Clark Fork. Here the water was turbid; visibility was poor. The rocks of the streambed were largely cemented together with silt, leaving no habitat for stoneflies or *Arctopsyche*. I didn't see a single fish. I didn't see a single insect. Some people claim that the Clark Fork today is actually much improved over its sorry condition two decades ago, before the Company adopted certain technical measures to mitigate the toxicity of its releases. Maybe those people are right. But I remain skeptical. The river I was swimming through, with my eyes open and my nose very close to the bottom, was definitely no basis for passing out congratulations.

This dramatic lack of vitality proves nothing, of course, about what causal role the smelter wastes and the erosion from denuded hillsides around Anaconda may or may not still be playing. It simply correlates. Consider it, if you wish, purest coincidence. It is not, however, mythical. It is real.

Later Benowitz and I were careful to shower ourselves down with clean water. "River-snorkeling," he told me, and he should know, "is not supposed to be a death sport."

The Tree People

SOME HUMANS HAVE A SPECIAL RELATIONSHIP with trees.

I'm thinking here not of the professional foresters, nor the academic dendrologists, certainly not the barrel-chested and flannel-shirted fallers. No, it's gentler folk I have in mind. Persons neither scientific nor pragmatic, whose encounters with trees tend to be more intimate, more spontaneous, marked by an altogether different degree of sensitivity and—it might not be going too far to insert the word "mutual" here—appreciation. People who can actually quiet themselves sufficiently to gaze at one individual tree and perceive there a real living creature conducting its own mortal business. This isn't as easy as it sounds. "The tree which moves some to tears of joy is in the eyes of others only a green thing which stands in the way," wrote William Blake. These genuine tree people are rare.

John Muir was one—read his account of riding out the thrills of a mountaintop storm while perched in the upper branches of a hundred-foot spruce. Another was that curious historical figure Jonathan Chapman, dead in 1845, later sentimentalized under the name Johnny Appleseed. Still another is the British novelist John Fowles, who has written an interesting and little-known nonfiction book titled *The Tree*, in which he avows: "If I cherish trees beyond all personal (and perhaps rather peculiar) need and liking of them, it is because of this, their natural correspondence with the greener, more mysterious processes of mind—and because they also seem to me the best, most revealing messengers to us from all nature, the nearest its heart." Among this group of tree-loving people, too, is my own father.

Unlike Muir, he has never rambled solitarily among the California sequoias, my father. Unlike Fowles, he offers no elaborate philosophical theories for his special attachment. Like Chapman, he is simply a planter. Ever since I can remember—let's say at least thirty years—this man has been planting and tending and doting on trees. He has never sold a board-foot of timber. He has never carried a bushel of fruit to a fair. He barely consents to own a saw. And behind him his life stretches out like a burgeoning, flourishing woodlot.

For a long time it made no particular sense to me.

At the beginning there was a half-acre of nearly bare real estate, formerly farmland, at the suburban fringe of a city in the southern Midwest, not far from the Mason-Dixon Line. Upon the half-acre sat a new house and a stately old black walnut tree, not much else; beyond the fence marking the west edge of the property rose a wild stand of high grass and thistle, through which a foot trail led to unspoiled hardwood forest that went on for several miles. Across that fence was a miniature wilderness area for the delectation of young boys. Then, gradually, bits of forest came to the half-acre lot.

A soft maple was planted up front near the road. A hard maple just out the kitchen window. A sweet gum beside the driveway. A pin oak near the old well. An apple tree off at the northwest corner, keeping company with the compost heap. A little dogwood. A Scotch pine, which often afterward seemed to be struggling against heat prostration. More maples along that west fence. Eventually, getting fancy, a ginkgo. And a magnolia tree, a hapless and delicate magnolia, to the right of the front door. The earliest of these were poached as saplings from the adjacent woods, carefully transported home on the footpath, and replanted as adoptees; later it became necessary to patronize a nursery. From my point of view (roughly waist-high then), the place had become a nursery itself. Finally the man in question bought me a rake, and I was not amused.

It had become necessary to patronize a nursery because by the time I was old enough to operate that new rake, the wilderness area over the back fence had disappeared. Bulldozers had scraped it away. In its place there was now a tract suburb of medium-sized boxes. Paved driveways and sidewalks. Tulips. Myrtle in neat patches. Precious few trees.

I sat high in the crow's nest of that old black walnut tree, years passing, and watched this transformation. A valuable early lesson, with the resonance of a parable. Still grudging my time at the rake, I could see after a while that the man, the planter, my father, was not insane after all. Not even perverse. As the forest was massacred, as the neighborhood turned into concrete and crabgrass, on our small island there was a continual spreading of new branches. Local trees and exotics thrown together in strange juxtapositions, most of them thriving, fighting one another genially for sunlight and water and the attentions of the chief arborist. He was running his own gene bank. I would not want to be so highfalutin as to call this half-acre a symphonic orchestration of trees; but it represented at least a pretty good Dixieland band. The planter enjoyed his son's approbation for a couple more years, until a midwinter ice storm hit the magnolia, at which point it seemed that things were perhaps being taken too far.

This would have been about 1963. In that peculiar borderland climate, ice storms were an occasional feature of what passed for winter. Cold sleet would begin falling at night, as the temperature dropped, and by morning the entire city would be glazed with a very beautiful and extremely treacherous eighth-of-an-inch layer of clear ice. Power lines down. Fender smashing against fender. Hips being fractured. I remember vividly how ice fragments and sparks would fly from the overhead wires that fed juice, via long antennas, to the old electric city buses. The ice storm in 1963 was especially bad, and instead of an eighth-inch thickness, there was a quarter inch. Over any large surface area, like the crown of a tree, that amounted to considerable weight. The conifers, adapted to serious snows, could take it. The hardwoods were bare and streamlined. But the magnolia, a southern-bred creature, too naive and injudicious to drop its big trowel-shaped leaves from its brittle limbs after the autumn frosts, was caught in a wretched position.

Every leaf was lacquered thickly with ice, hundreds of pounds in all, the whole tree about to collapse like so much Steuben glass in a garbage compacter. And so, on that Saturday morning, there was the droll spectacle of a man and his fifteen-year-old son (the latter an unwilling draftee, with places to go and other enormously pressing things, now forgotten, to do) taking turns on a stepladder to break the ice—gingerly, one leaf at a time, with raw cold hands—off that desperate magnolia.

I remembered the magnolia's trauma, and its rescue, just yesterday while reading some scientific papers on an extraordinary species of tree called the bristlecone pine. The bristlecone was originally to be the main subject of this essay—now largely preempted, but never mind. The bristlecone will get its full starring role some other time. It can wait. It knows how. It is a tree accustomed to taking the long view. Unrecognized by biologists until about thirty years ago, bristlecone pines in the mountains of the American Southwest are today known to be among the oldest living creatures on Earth.

We're not talking about the age of a species, understand, but about venerable individuals. One noble specimen of bristlecone, found alive in 1964 on the shoulder of a high peak in eastern Nevada, was calculated to be 4,900 years old. That single tree sprouted from its seed and began putting out needles, in other words, around the same time the Egyptians established their first kingdom. No pyramids yet. The book of Genesis was still in galleys. This was a very old tree.

A number of curious facts emerged from those journal papers on the bristlecone, most of which have no pertinence here, no bearing upon the subject of the man, the ice storm, and the magnolia. But two of them do.

First: Dendrologists have discovered that longevity among individuals of this most long-lived of earthly creatures is inversely related to the hospitableness of its living conditions. The tree thrives, over great lengths of time, on adversity. The more harsh and ungiving its particular locale, the longer a bristlecone tends to live. At lower elevations within the mountains of its native range, places where soil is decent, wind and erosion are not extreme, water is available in good supply, the bristlecone grows large and robust, but does not seem to survive much beyond 1,500 years. These ages can be gauged rather precisely, from a core sample or a full cross-section, by counting the annual rings laid down through the trunk. At higher elevations, right up at the edge of the timber-line, on steep south-facing slopes of stony soil that is poor in organic material and chemical nutrients, where little water is available, winds are relentless, growing conditions are generally lousy—here the bristlecone lives as a gnarled dwarf. But long. Often enough in this harshest environment a bristlecone survives its four thousandth birthday. Obvious moral: When the growing is tough, the tough keep growing.

The second odd fact is corollary to that one. Certain dendrologists who count bristlecone tree rings have taken to writing of two separate characters of tree within the species. These are the "sensitive" bristlecones and the "complacent" bristlecones. The sensitive bristlecones are those that respond to climatic fluctuations, such as a year of exceptional drought, by laying down a drastically narrower growth ring that year, or possibly no ring at all. A complacent tree records no such response. Maybe it shouldn't be surprising that the complacent trees (as reported in *Science*, 1968) tend to be those that are younger and more comfortable. Meanwhile the ancient trees, struggling through four thousand years of thirst and starvation, solitary on exposed ridges, grotesquely shaped, hunkered down, clinging to life with only two or three green branches—these are the ones that leave a record of sensitivity.

Now it seems to me that this discernment of "sensitivity" and "complacency" among various individual pine trees, silently living to millennial ages on their high mountain slopes, must constitute some sort of breakthrough percipience for the botanical sciences. And it compels me to wonder about that beleaguered magnolia on the half-acre woodlot at the edge of the midwestern city.

Was the magnolia by disposition a complacent creature? Or was it, as I hope, sensitive? Did it appreciate, in some sense, during the year of growth following that 1963 ice storm, what the man with the stepladder and the raw hands had done for it?

The house and the half-acre have long since been sold to

strangers. The magnolia, last time I was in the city and drove past, looked neglected: a broken crown, whole branches on which the leaves were a sickly brown. Too bad. It may not survive another big ice storm. Certainly it won't live to see four thousand.

But might there remain, perhaps, in its heartwood, some record of response—just a slight thickening to one annual ring, like a grateful sigh—to that mildly eccentric act of love? Can a magnolia remember a man?

ELOQUENT PRACTICES, NATURAL ACTS

Love's Martyrs

LIKE WOODY ALLEN, the English poet John Donne was obsessed in his younger days by love and death. Throughout Donne's early work those two motifs recur again and again, linked so closely together that they come to seem almost logically inseparable, two sides of a macabre equation, evoking each other almost interchangeably. Love is a manner of dying, Donne suggests; and vice versa. For instance: "I cannot say I lov'd, for who can say / Hee was kill'd yesterday?" Another poem contains the tender sentiment "Since thou and I sigh one another's breath, / Who e'r sighes most, is cruelest, and hastes the other's death." In still another, a man speaking from his own grave declares himself "love's martyr," dead of an excess of passion. During the sixteenth century in England, this oxymoronic linkage of the two concepts, love and death, was enough to get an ardent young man eventually categorized as one of the "metaphysical" poets. Nowadays it would make him a theoretical population ecologist.

The notion with which John Donne was flirting, in all that love-and-death poetry, is studied today as a curious phenomenon of evolutionary ecology and denoted by the label *semelparity*. That fancy word is, of course, just another bit of the formal jargon that scientists take cruel joy in inventing. The same thing is more casually known, with a lewd nod from one science to another, as Big Bang reproduction.

Semelparity: An animal or a plant waits a very long time to pro-create only once, does so with suicidal strenuousness, and then promptly dies. The act of sexual reproduction proves to be ecstati-cally fatal, fatally ecstatic. And the rest of us are left merely to say, Wow.

As a strategy for perpetuation of a species, semelparity is not well understood. But the list of known semelparous creatures is intriguingly diverse. Bamboos do it. A group of hardy desert plants called the agaves do it. Pacific salmon do it. The question is why. What can these three kinds of organisms, apparently so dissimilar, have in common? Why should all three, living in drastically differ-ent environments with drastically different life histories, be simi-larly committed to dying for one taste of love?

The answer, according to cautious speculation by some ecolo-gists, may be as simple as a few symbols. Natural selection, these researchers say, tends to maximize, for each age i, the sum $B_i + p_i(v_{i+1}/v_o)$. They may be right, but if you're like me, any such clot of algebraic erudition immediately causes alarm bells to ring in your head, sprinkler systems to begin dousing your overheated brain, and your eyes to slide straight off the page like a cheap ballpoint skidding across oilpaper. But wait. The idea wrapped in that ugly cryptogram happens to be rather interesting, and just possibly it can also be said in English.

First, a few concrete facts. Five different species of salmon migrate regularly from the Pacific Ocean into the rivers of western North America. They are headed upstream to spawn, and for each indi-vidual fish the journey is a return trip back to that same freshwater tributary where it began its life. Some of these salmon (the Chi-nook of the Yukon River, for instance) will travel as much as 2,000 miles, climbing through rapids, making ten-foot leaps to clear the cascades, dodging predators, fighting constantly upriver at an unflagging pace of perhaps 50 miles per day. The effort and deter-mination involved are prodigious, but it is a one-way trip. Soon after having spawned, every male and every female of these five species is dead. Decomposing corpses pile up in the eddies, turn-ing clear mountain water funky with rotting flesh. Among those lucky fish that have completed the trip, won a mate, found genetic fulfillment in the gentle current above a gravel spawning nest, there are no survivors.

Certain types of bamboo make a long journey to breed also, but the distance they cover is measured in time. The common Chinese species *Phyllostachys bambusoides*, for instance, has a regular life cycle of 120 years. Historical sources back into the tenth century record its episodes of massive synchronous flowering. Each time around, a vast number of *P. bambusoides* individuals begin life together as new sprouts; for 120 years they grow taller and sturdier, putting out leaves and branches, storing away energy, adding clones of themselves to the population by a nonsexual budding process, maturing together into a dense grove; then, after the appointed twelve decades, they suddenly and simultaneously produce an awesome profusion of flowers. The blossoms fertilize one another by wind. Seeds fall like heavy hail, coating the ground, ankle-deep to a man. And the 120-year-old progenitors all immediately die.

Members of the plant genus *Agave*, meanwhile, are content through long years of growth and celibacy to resemble giant artichokes peeking out between desert rocks. They are succulents, related to lilies but preferring a hard life on dry hillsides, such as those of the Sonoran Desert. They unfold their leaves in a spiral rosette, each leaf tipped with a sharp spine for protection against browsers, and they range by species from the size of a small porcupine to the size of a 300-pound octopus. Like the bamboo, they can set clones of themselves by budding, but their primary mode of reproduction is sexual. Years go by uneventfully, an agave grows large and vigorous until, with startling abandon, one season it produces a towering flower stalk. This great inflorescence will be notable not for the beauty or fragrance of its blossoms but for its sheer height. From a big agave, in just three or four months the stalk might shoot up thirty feet, sturdy and straight as a young lodgepole pine. It's a flower that you'd cut with a chainsaw. Pollination is done by insects and bats; seeds ripen and drop. And down below, shriveled and brown, the agave dies away, as though speared through the heart by its own flowery stalk.

This is semelparity, at work in the moist tropics of Asia, on the sere slopes of southern Arizona, at the headwaters of the Yukon River. A curious person naturally wants to know why, in each case, a single act of sex should prove deadly. But even more interesting, it seems to me, is the first derivative of that question: Is there any *one* answer that can explain why sex is terminal in *three* such disparate cases?

A man named William M. Schaffer says yes, and offers this:

$$\Sigma_i \, [B_i + p_i(v_{i+1}/v_0)] = k$$

Hold the sprinklers, hold the alarm. What he is talking about is simply a delicate balance between love and death.

Dr. Schaffer is a respected theoretical ecologist at the University of Arizona. In a half-dozen papers published over the past decade, alone and with various coauthors, he has proposed a theory of how evolution tends to produce optimal life-history strategies for different animals and plants. This theory entails enough mathematical bebop to make the average human swoon. Descriptive numbers for a particular species can be inserted into Schaffer's equations, a lever is then pulled, a crank is turned, and the theory will posit how the creature should act if it wants to survive the long-term Darwinian struggle. That's the kind of thing theoretical ecologists do. Semelparity is quite useful in such theorizing, because it involves the love-and-death balance taken to one logical extreme.

Dr. Schaffer calls that balance "the trade-off function." The particular trade-off at issue is between present and future—that is, the immediate prospect of producing offspring versus the chance of surviving to produce other broods later. An underlying premise here is that an animal or a plant has, at each stage in its life, only a limited total amount of available energy that it can spend on the business of living. That limited energy must be shared out among three fundamental categories of effort—routine metabolism, growth, and reproduction—and an optimal life history is one that balances the shares most efficiently to produce evolutionary success. Evolutionary success, of course, is measured on one simple scale: How many offspring does the creature leave behind? All this is quite basic.

Schaffer's trade-off equation merely codifies that crucial balance between present and future, between short-term and long-term concerns, between effort devoted to immediate self-preservation and effort devoted to parenthood. His B_i represents the reproductive potential of a given creature right now. The factor p_i stands for the probability (or improbability) that the given creature, having bred once, will survive to breed again later. The parenthesis (v_{i+1}/v_o) will be the remaining reproductive potential that an experienced parent can expect still to possess at that hypothetical later time. Balance all these considerations against one another, with just the proper commitment of energy (at each age) to metabolism, to growth, to reproduction, and the result is high evolutionary success. This is the burden of $\Sigma_i [B_i + p_i(v_{i+1}/v_o)] = k$.

Now let's try it in English. If a female Chinook salmon, having swum 2,000 miles up the Yukon River, having climbed rapids and dodged otters and leapt cascades, having beaten her fins to tatters digging a gravel nest and fought off other salmon to keep the spot inviolate—if this poor haggard creature has virtually no chance of surviving to accomplish the same entire feat again, then she will be required by the forces of natural selection to sacrifice herself totally, in one great suicidal effort of unstinting motherhood from which she cannot possibly recover. She will lay about five thousand eggs. And then she will croak.

Likewise for those Sonoran agaves, for those Chinese bamboos. The terms of the trade-off are the same, the results are the same, only the numbers and the details are different. Bamboos seem to sacrifice themselves for the sake of predator satiation—that is, producing so many seeds that after all the rats and jungle fowl of China have eaten their fill, a few seeds will still be left to germinate. The agaves compete with each other to produce taller and yet taller flowers, apparently because their pollinators deign to visit only the tallest. The evolutionary consequence in each case, as with salmon, is semelparity.

But Dr. Schaffer's neat mathematical model is not without gaps, not without weaknesses. What about the *Atlantic* salmon, for instance, which faces an almost identical set of circumstances to those the Pacific species do yet which doesn't resort to semel-parity? It can be argued that Schaffer's equations constitute an unduly abstract version of reality. To some observers, such airy theorizing has little more connection to the untidy actualities of ecological fieldwork than it does to, say, metaphysical poetry.

At Christmas of the year 1600, John Donne secretly married a young girl named Anne More, a gentle but dignified sixteen-year-old whose hotheaded father was a powerful nobleman, serving as Queen Elizabeth's Lieutenant of the Tower. Donne himself was twenty-seven and employed as private secretary to the girl's uncle. It was a reckless move, this marriage, putting passion before prudence, and Donne knew that. He suffered consequences: dismissed from his job, briefly imprisoned, denied the dowry that Anne otherwise would have brought, and left to struggle for years on the margins of poverty with his adored wife and their many children. Around that time he wrote:

> *Love with excesse of heat, more yong than old,*
> *Death kills with too much cold . . .*
> *Once I lov'd and dy'd; and am now become*
> *Mine Epitaph and Tombe.*
> *Here dead men speake their last, and so do I;*
> *Love-slaine, loe, here I lye.*

It's highly unlikely that John Donne ever set eyes upon a Pacific salmon, or a Sonoran agave, or even a transplanted grove of *Phyllostachys bambusoides* in some London botanical garden. But we can safely assume that he would have understood.

A Deathly Chill

DEATH IS PERSONAL. It's sealed over by subjectivity and silence; one of those things either you do or you talk about, but not both. Notwithstanding the work of Elisabeth Kübler-Ross, it is the single inevitable human enterprise that we can have no hope of comprehending in advance. The contemplation of death is, after all, something live people engage in, by way of linguistic bamboozlement and philosophical placation of themselves. I repeat all this obvious stuff here because of the Ram Patrol of Chattaroy, Washington, and the question of hypothermia.

On August 10, 1982, an Associated Press item ran in a corner of the back page of my local newspaper under the headline "Hypothermia Blamed in Deaths of Scouts." The story was bizarre and pathetic. Four young Boy Scouts and two adult leaders had been found drifting, dead, in a cove of a glacier-fed Canadian lake near the west border of Banff National Park. They were stragglers from a canoe trip that had included twenty-three other boys and men, and they had been missing since just the previous afternoon. When discovered, all of the corpses were floating, head up and neatly strapped into lifejackets, not far from their undamaged canoes. There was no sign of accident, desperate struggle, or panic. One of the two adults was still wearing his glasses and hat. "They were in the Ram Patrol, our most experienced group," a scoutmaster told the AP. "We had them follow the others because they were the best." The water temperature was steady at around 45°F, and all of the scouts had been in and out of it, swimming and bathing, for the whole week. The presumed cause of death was hypothermia.

It would not be quite accurate to say that the four boys and two men of the Ram Patrol had frozen to death—not at 45 degrees Fahrenheit. Rather, evidently, they had been *chilled* to death. "It was like they had just gone to sleep in the water," said the scoutmaster. "They probably ran out of energy and died." What seemed obvious to this man, by hindsight, had apparently stolen upon six robust campers like a Mosaic angel of death. Given the nature of hypothermia and the nature of water, it is not hard to believe.

Seventy years ago, hypothermia, like radiation sickness, was unheard of. People in those days died of consumption, yellow fever, childbirth, the flu. They also, in cases of mishap on the high seas, died of "drowning." But drowning was merely the standard official and popular presumption, clung to for lack of a better one— live people doing their best, again, to get a grip on the lonely and personal business of death. As late as 1969, a physiologist from Oxford University, W. R. Keatinge, wrote that "until recently even experts commonly regarded drowning as the only important hazard to life in the water. Those who did look forward seldom appreciated any other hazards except thirst and attack by sharks. This belief is still common. It is almost routine for anyone who dies in the water to be said to have drowned, not only in everyday conversation and the press but often in official reports." For example, the case of the *Titanic*.

It was on its maiden trip when an iceberg hit the ship. This was shortly before midnight on April 14, 1912, a chilly spring night in the North Atlantic, about 400 miles off the coast of Newfoundland, and the temperature of the water in which the *Titanic* sank, so quickly, was hovering around 32°F. During the early minutes of pandemonium, roughly a third of the passengers and crew managed to get safely aboard lifeboats, either dry-shod from the side of the ship itself or after a brief dunking. The other 1,489 people were left swimming, but there were far more than enough lifejackets to take care of everybody. Within just one hour and fifty minutes another ship, the *Carpathia*, arrived on the scene to begin scooping up survivors. Now the shocking part. The *Carpathia* was able to save almost all of those lucky or assertive folks who had gotten themselves places in the lifeboats; *every one* of the other 1,489 people, most of them bobbing there nearby in perfectly decent lifejackets, was dead.

Afterward an official report came down from the superintendent of the Port of Southampton, under the title "Particulars Relating to the Deaths of Members of the Crew Lately on Board the S.S. *Titanic*." This document included a roll of fatalities that ran nineteen pages, and after each name the cause of death was cited as drowning. In all the various investigations and reports following the *Titanic* disaster, according to Keatinge, there was hardly a mention of immersion hypothermia—under that phrase or any other—as a cause of or a contributing factor in death.

Nowadays scientists and maritime people know better. Perhaps no form of exposure to nature's brutal indifference is deadlier and (aside from running afoul of a grizzly or toppling off El Capitan) swifter than hypothermia. Yet it is also insidiously subtle. About six hundred Americans die from it each year, and despite the common notion that associates hypothermia with arctic cold, most of those six hundred victims were never in any danger of frostbite. Many of them were subjected only to cool or even mild temperatures, in the forties and fifties, but for one reason or another they got caught with wet clothing in the path of brisk winds and couldn't protect themselves until too late. Many others were simply plunked into an ocean or a stream or a lake, like the Ram Patrol, and for one reason or another couldn't get out. They didn't freeze to death, like that smug dude in the Jack London story "To Build a Fire," which still represents the most widely known, and misleading, depiction of fatal hypothermia. They chilled to death. It doesn't take long.

Most old-fashioned clothing (wool, of course, being the exception) loses up to 90 percent of its insulating value when it is wet. Put a rain-drenched mountaineer on a breeze-raked ridge and, without a windbreaker, his life will be in jeopardy within half an hour. But full immersion in water, an occupational hazard of sea travelers since the time of Noah, is deadlier still, because the thermal conductivity of water is 240 times that of air. While a man overboard sculls gently to keep his face out of the waves, rides without further effort in his lifejacket, waits hopefully for speedy rescue, the water sucks heat—and therefore life—from the core of his body at a merciless rate. Immersed in water at 32°F, like the *Titanic* passengers, the average human will die within an hour. Immersed at 59°F, he will die after six hours. And 59°F happens to be warmer than practically all of the coastal and inland waters of North America; in fact, it is warmer than most of the surface water on the planet ever gets (outside of the tropics) through an entire year.

No wonder shipwreck, grand and small, has killed so many good swimmers. No wonder Madame Sosostris, famous clairvoyant, wisest woman in Europe, said: "Fear death by water."

As a body begins losing large amounts of heat to the surrounding environment, two things happen. The less serious is frostbite, in which blood circulation to the extremities is automatically reduced as a desperate measure to conserve heat; this results in a drastic differential between the temperature of the skin and the temperature of the thoracic interior, and those expendable fingers and toes are sacrificed to maintain thermal stability in the body's vital core. The more serious phenomenon is hypothermia, occurring when no such differential, no such desperate sacrifice, can prevent the core temperature itself from plummeting. As the core temperature falls, the symptoms of hypothermic trauma develop in progressive stages. A physician and mountaineer named Ted Lathrop, in a pamphlet published by the Mazamas climbing club, has described those stages in detail.

Dropping from a normal 98.6° down to 96° at the body core, says Lathrop, the victim will show uncontrollable shivering and a distinct onset of clumsiness. From 95° down to 91° the shivering will continue, and now speech will become slurred; mental acuity will decrease; there may also be amnesia. During this stage often come those crucial mistakes in judgment that prevent a victim from taking certain obvious steps that could save him from death. Between 90° and 86° the shivering will be replaced, says Lathrop, with extreme muscular rigidity, and exposed skin will sometimes appear blue or puffy. Mental coherence may be negligible, and amnesia may be total, though the person may still be able to walk, and if he is unlucky, his companions may not yet have noticed that he is in serious trouble. Down around 81° he will slide into a stupor, with reduced rates of pulse and respiration. Two or three degrees colder than that at the body core and he goes unconscious. His heartbeat may now be erratic, yet even if it remains steady, there is a grave problem: Human blood cooled to this temperature becomes reluctant to turn loose the oxygen it's supposed to transport, so despite continued circulation, the brain and the heart muscle may be starved of the oxygen they require. If the core temperature falls further, below 78°, those brain centers that govern heartbeat and respiration will probably give out. There will be cardiac fibrillation—that is, the heart will be gripped with disorganized spasmodic twitching. Also now, pulmonary edema and hemorrhage—the lungs suddenly filling with clear cellular liquids and blood. The person may vomit or just cough rackingly, a bit of pink foam frothing out from between his lips. And then he is dead.

The coldest spot of skin on his coldest toe may be no colder than 59°F. But his core has fallen to 75°, and all the gears have seized.

The conditions in Lake McNaughton, British Columbia, where the Ram Patrol came to their end, were more than sufficiently inhospitable to bring on this sequence of stages and to justify that surviving scoutmaster in his diagnosis. True enough, it would be hard to imagine how six healthy young men could drown under such circumstances, but not hard at all to figure how hypothermia might have killed them. The lake had gotten a bit choppy in late afternoon, and the Ram Patrol must have pulled into that cove for shelter, the scoutmaster guessed, when their canoes started taking on water. They seem to have climbed out of their boats in the shallows, he guessed, and even succeeded in dumping both canoes empty and righting them again. Or something. "Then they must have run out of energy and hypothermia set in," said the scoutmaster to the Associated Press.

But then later that week I made a call to the coroner of Revelstoke, British Columbia, within whose jurisdiction the Lake McNaughton incident occurred. The coroner told me something different. Sometime after the first newspaper story, autopsies were performed. Hypothermia, it had been quietly concluded, was *not* the cause of the deaths.

The lungs of the victims were filled not with blood and clear bodily fluids but with lake water. The four boys and two men of the Ram Patrol, floating in lifejackets near their empty canoes, had all drowned.

No one knows how. No one is likely to find out. Early symptoms of hypothermia, such as lassitude or muscular rigidity, may have made some secondary contribution, said the coroner. But that, he admitted, was purely speculative. With no witnesses and no survivors, the truth could only be guessed at. Death is personal.

Is Sex Necessary?

BIRDS DO IT, BEES DO IT, or so goes the tune. But the songsters, as usual, would mislead us with drastic oversimplifications. The full biological truth happens to be more eccentrically nonlibidinous. Sometimes they *don't* do it, those very creatures, and get the reproductive results anyway. Bees of all species, for instance, are notable for their ability to produce offspring while doing *without*. Birds mostly do mate, yes, but at least one variety—the Beltsville Small White Turkey, a domestic breed out of Beltsville, Maryland—has achieved scientific renown for a similar feat. What we're talking about here is celibate motherhood, procreation without copulation, a phenomenon that goes by the name *parthenogenesis*. Translated from the Greek roots: virgin birth.

And you don't have to be Catholic to believe in this one.

Miraculous as it may seem, parthenogenesis is actually rather common throughout nature, practiced regularly or intermittently by at least some species within almost every group of animals except (for reasons still unknown) dragonflies and mammals. Reproduction by virgin females has been discovered among fishes, amphibians, birds, reptiles,* crustaceans, mollusks, ticks, the jelly-

---

* Within recent months a female Komodo dragon named Flora, at a zoo in England, produced five hatchlings despite her total lack of breeding contact with any male. Roughly seventy other reptile species besides *Varanus komodoensis* are now known to be capable of parthenogenesis. Somehow, though for no particular reason, it seems even

fish clan, flatworms, roundworms, segmented worms; and among insects (notwithstanding those unrelentingly sexy dragonflies) it is especially favored. The order Hymenoptera, including all bees and wasps, is uniformly parthenogenetic in the manner by which males are produced: Every male honeybee is born without any genetic contribution from a father. Among the beetles, there are thirty-five different forms of parthenogenetic weevil. The African weaver ant employs parthenogenesis, as do twenty-three species of fruit fly and at least one kind of roach. Gall midges of the species *Miastor metraloas* are notorious for the exceptionally bizarre and grisly scenario that allows their fatherless young to see daylight: *M. metraloas* daughters cannibalize the mother from inside, with ruthless impatience, until her hollowed skin splits open like the door of an overcrowded nursery. But the foremost practitioners of virgin birth—their elaborate and versatile proficiency unmatched in the animal kingdom—are undoubtedly the aphids.

Now no sensible reader, not even one who has chosen *this* book, can be expected to care much, I realize, about aphid biology qua aphid biology. But there's a larger reason for dragging you into the subject. The life cycle of these nebbishy insects, the very same that infest rosebushes and houseplants, exemplifies not only how parthenogenesis works but also, very clearly, why evolution has devised such a reproductive shortcut.

First the basics. A typical aphid, which feeds entirely on plant juices tapped from the vascular system of young leaves, spends winter as an egg, dormant and protected. The egg is attached near a bud site on the new growth of, say, a poplar tree. In March, when the tree sap has begun to rise and the buds have begun to burgeon, the egg opens and an aphid hatchling appears, promptly plugging its sharp snout into the tree's tender plumbing. This solitary individual aphid will be, necessarily, a wingless female. If she is lucky, she will become sole founder of a vast aphid population. Having

---

more amazing in a vertebrate—such as a giant lizard or a bird—than in aphids. Probably that's because we are vertebrates ourselves, and biased toward believing that sexual reproduction is a "higher" form than asexual.

sucked enough poplar sap to reach maturity, she produces (by live birth now, not egg-laying, and without benefit of a mate) daughters identical to herself. These wingless daughters also plug into the tree's flow of sap, and they also produce wingless daughters—whose daughters produce more daughters, geometrically more, generation following generation until sometime in late spring, when crowding becomes an issue and that particular branch of that particular tree can support no more thirsty aphids. Suddenly there is a change: The next generation of daughters are born with wings. They fly off in search of a better situation.

One such aviatrix lands on a herbaceous plant—a young climbing bean, say, in someone's garden—and the pattern repeats. She plugs into the sap ducts on the underside of a new leaf, commences feasting, robbing the plant of its vital juices, and then delivers by parthenogenesis a great brood of wingless daughters. The daughters beget more daughters, those daughters beget still more, and so on, until the poor bean plant is encrusted with a dense population of these fat little sisters. Then again, neatly triggered by the crowded conditions, a generation of daughters are born with wings. Away they fly, looking for prospects, and one of them lights on, say, a sugar beet. (The switch from bean to beet is possible for our species of typical aphid, because it is not a dietary specialist committed to only one plant.) The sugar beet before long is covered, sucked upon mercilessly, victimized by a horde of mothers and nieces and granddaughters. Still not a single male aphid has appeared anywhere in the lineage.

The lurching from one plant to another continues; the alternation between wingless and winged daughters continues. Then, in September, with fresh and tender plant growth increasingly hard to find, there comes another change.

Flying daughters are born who have a different destiny: They wing back to the poplar tree, where they give birth to a crop of wingless females unlike any so far. These latest girls know the meaning of sex! Meanwhile, at long last, the starving survivors back on that final bedraggled sugar beet have brought forth a generation of males. The males too have wings. They take to the air in search of poplar trees and first love. *Et voilà.* The mated females lay eggs that will wait out the winter near bud sites on that poplar tree, and the circle is closed. One single aphid hatchling—call her the matriarch—in this way can give rise in the course of a year, from her own ovaries exclusively, to roughly a zillion aphids.

Good for her, you say. But what's the point of it?

The point, for aphids as for most other parthenogenetic animals, is 1) exceptionally fast reproduction that allows 2) maximal exploitation of temporary resource abundance and unstable environmental conditions, while 3) facilitating the successful colonization of unfamiliar habitats. In other words, the aphid, like the gall midge and the weaver ant and the rest of their fellow parthenogens, is by its evolved character a hasty opportunist.

This is a term of science, not of abuse. Population ecologists make an illuminating distinction between what they label "equilibrium species" and "opportunistic species." According to William Birky and John Gilbert, from a paper in the journal *American Zoologist*: "Equilibrium species, exemplified by many vertebrates, maintain relatively constant population sizes, in part by being adapted to reproduce, at least slowly, in most of the environmental conditions which they meet. Opportunistic species, on the other hand, show extreme population fluctuations; they are adapted to reproduce only in a relatively narrow range of conditions, but make up for this by reproducing extremely rapidly in favorable circumstances. At least in some cases, opportunistic organisms can also be categorized as colonizing organisms." Birky and Gilbert emphasize that the potential for such rapid reproduction is "the essential evolutionary ticket for entry into the opportunistic life style."

And parthenogenesis, in turn, is the greatest time-saving trick in the history of animal reproduction. No hours or days are wasted while a female looks for a mate; no minutes lost to the act of mating itself. The female aphid attains sexual maturity and, bang, she becomes automatically pregnant. No waiting, no courtship, no fooling around. She delivers her brood of daughters, they grow to puberty and, zap, another generation immediately. The time saved by a parthenogenetic species may seem trivial, but it is not. It adds up dizzyingly: In the same duration required for a sexually reproducing insect to complete three generations for a total of 1,200 offspring, an aphid can progress through six generations (assuming the same maturation rate and the same number of progeny per litter) to yield an extended family of 318,000,000.

Even this isn't speedy enough for some restless opportunists. That matricidal gall midge *Miastor metraloas*, whose larvae feed on fleeting eruptions of fungus under the bark of trees, has developed a startling way to cut further time from the cycle of procreation. Far from waiting for a mate, *M. metraloas* does not even wait for maturity. When food is abundant, it is the *larva*, not the adult female fly, who is eaten alive from inside by her own daughters. And as those voracious daughters burst free of the husk that was their mother, each of them already contains further larval daughters taking shape ominously within its own ovaries. While the food lasts, while opportunity endures, no *Miastor metraloas* female can live to adulthood without dying of motherhood.

The implicit principle behind all this nonsexual reproduction, all this hurry, is simple: Don't pause to fix what isn't broken. Don't tinker with a genetic blueprint that works. Unmated female aphids, and gall midges, pass on their own genotypes virtually unaltered (except for the occasional mutation) to their daughters. Sexual reproduction, on the other hand, exists to allow genetic change. The whole purpose of joining sperm with egg is to shuffle the genes of both parents and come up with a new combination that might perhaps be more advantageous. Give the kid some potent new mix of possibilities, based on a fortuitous selection from what Mom and Pop individually had. Parthenogenetic species, during their hurried phases at least, dispense with this genetic shuffle. They stick stubbornly to the genotype that seems to be working. They produce (with certain complicated exceptions) natural clones of themselves.

But what they gain thereby in reproductive rate, in great explosions of population, they lose in flexibility. They minimize their genetic variability—that is, their options. They lessen their chances of adapting to unforeseen changes of circumstance.

Which is why more than one biologist has drawn the same conclusion as M.J.D. White: "Parthenogenetic forms seem to be frequently successful in the particular ecological niche which they occupy, but sooner or later the inherent disadvantages of their genetic systems must be expected to lead to a lack of adaptability, followed by eventual extinction, or perhaps in some cases by a return to sexuality."

So it *is* necessary, even for aphids, this thing called sex. At least intermittently. A hedge against change and oblivion. As you and I knew it must be. Otherwise, surely by now we mammals and dragonflies would have come up with something more dignified.

Desert Sanitaire

THE ENGLISHMAN T. E. LAWRENCE, he of Arabian fame, was supposedly once asked why he so loved the desert. Skeptical historians have treated Lawrence far less kindly than Peter O'Toole and David Lean did, suggesting that far from being the reluctant demigod and charismatic catalyst of Arab revolt, as so appealingly pictured, he was more on the lines of a conniving, ambitious, and perpetually mendacious poseur. Also, unlike O'Toole, he stood only five foot three. To hear it from some of his biographers, the truth was not in Lawrence. He colluded with Lowell Thomas (in those days a young showman, more interested in dashing romance than journalistic fact) toward inventing and spreading the "Lawrence of Arabia" legend. He was a sadomasochistic neurotic whose entire life, say the critics, was "an enacted lie." He liked costumes and he invented his own heroism.

This revisionist view, sound in principle, may in fact be a little too harsh. Lawrence certainly had *something*, and maybe that something was almost as valuable as the habit of veracity or full mental health. He had panache. He had high style. He had the gift for capturing, if not strict autobiographical truth, at least the human imagination. *Why do you love the desert?* they asked him later, when he was languishing through his self-imposed obscurity back in soggy, dreary England. Reportedly he said: "Because it's clean."

At least, I hope he did.

Of course, by a literal reading the notion is nonsense. Clean of what, dirt? Not if dust and perspiration and a week's funky unwashed body grime can be counted. Clean of microbial infestation and many-legged vermin? Hardly. Perhaps clean of *human* infestation? That's more plausible as a guess of what he meant, given the misanthropic side of his disposition. Anyhow, if you have ever spent time out there—not in Arabia, necessarily, but in the desert—down on the very ground, crunching off the miles with your boots, maybe you understand something of what poor troubled Lawrence was getting at.

It's clean. It's austere. It's ascetic. Harshly infertile and fatally inhospitable. Solitary. Unconnected. It's notable chiefly for what it lacks. America's own preeminent desert anchorite seems to agree: Wherever his head and feet may go, says Ed Abbey, his heart and guts linger loyally "here on the clean, true, comfortable rock, under the black sun of God's forsaken country." It's clean.

But what *is* it, this thing of such noteworthy cleanliness? "There is no single criterion," according to the renowned desert botanist Forrest Shreve, "by which a desert may be recognized and defined." Still, we have to start somewhere. And a desert is one of those entities, like virginity and sans serif typeface, of which the definition must begin with negatives.

In this case, lack of water. Not enough rain. Less than ten inches of precipitation through the average year. A desert is not, most essentially, a hot place or a sandy place or a place filled with reptiles and cacti and dark-skinned people wearing strange headgear. Fact number one is that it's a dry place. Joseph Wood Krutch has written that "in desert country everything from the color of a mouse or the shape of a leaf up to the largest features of the mountains themselves is more likely than not to have the same explanation: dryness." From such a simple starting point, things get more complicated immediately.

The matter of sheer dryness, for instance, is less crucial than the matter of *aridity*, which is a measure of how much or how little water remains available on a particular landscape surface for how long. Ten inches of rain distributed evenly throughout a lengthy cool season will support plants and animals in modest profusion; ten inches dumped from a great cloudburst on one summer afternoon, then not another drop for the rest of the year, will produce a few hours of wild flooding and leave behind a typical parched desert, with wide empty arroyos and a scattering of peculiarly specialized creatures. Whatever water there may be comes and goes quickly in a desert, erratically, never remaining available over time. It abides not. It pours off the slopes of treeless mountains. It gathers volume in drywashes and roars peremptorily away. It soaks down fast through the sandy soil and is gone. Most of all, it evaporates.

That's the other prerequisite for any desert environment, lesser partner to dryness: evaporation, as wrought by heat and wind. A little rain falls occasionally, yes, but coming as it does in prodigal storms during the warmest months, burned off by direct sunshine and sucked away by the winds, the stuff disappears again almost at once. A system of land classification devised by Vladimir Köppen takes this into account, with a mathematical formula by which temperature and precipitation are together converted to an index of aridity. According to the Köppen method, any region where potential evaporation exceeds actual precipitation by a certain margin can be considered a desert. This rules out frigid locales with scant annual precipitation but plenty of permanent ice, such as Antarctica. Most of our own Southwest qualifies resoundingly.

But what, in the first place, makes a spot like Death Valley or Organ Pipe Monument so all-fired dry? Or a huge region like the Sahara? Or the Kalahari? Or the Taklimakan Desert of western China? Is it purely fortuitous that one geographical area—say, the Amazon basin—should receive buckets of moisture while another area not far away—the Atacama Desert in northern Chile—gets so little? The answer to that is no: not at all fortuitous. Three different geophysical factors combine, generally at least two in each case, to produce the world's various zones of drastic and permanent drought: 1) high-pressure systems of air in the horse latitudes, 2) shadowing mountains, and 3) cool ocean currents. Together those three cast a tidy pattern, north and south, girdling our planet with deserts like a fat woman in a hot red bikini.

Don't take my word for this: look at a globe. Spin it and follow the Tropic of Cancer with your finger as it passes through, or very near, every great desert of the northern hemisphere: the Sahara, the Arabian, the Turkestan, the Dasht-i-Lut of Iran, the Thar of India, the Taklimakan, the Gobi, and back around to the coast of Baja. Now spin again and trace the Tropic of Capricorn, circling down there below the equator: through the Namib and the Kalahari in southwestern Africa, straight across to the big desert that constitutes central Australia, on around again to the Atacama and the Monte-Patagonian of South America. This arrangement is no coincidence. It's a result, first, of that high-pressure air in the horse latitudes.

The horse latitudes (traditionally so called for tenuous and uninteresting reasons) encircle Earth in a pair of wide bands, one north of the equator and one south, along those two lines, Cancer and Capricorn. The northern band spans roughly the area between latitudes 20° N and 35° N, and the counterpart covers a similar area of southern latitudes. Between the two bands is that zone loosely called "the tropics," very hot and very wet, where most rainforest is located. This is also the zone of terrestrial surface that—because of its distance from the poles of rotation—is moving with greatest velocity as our planet spins through space. (The equator rolls around at better than 1,000 mph, while a point near the North Pole travels much slower.) For physical reasons only slightly less obscure than Thomistic metaphysics, the difference in surface velocity produces trade winds, variations in barometric pressure, and a consistent trend of rising air over the tropics. As the air rises, it grows cooler, therefore releasing its moisture (as cooling air always does) in generous deluge upon the tropical rainforests. Now those air systems are high and dry: far aloft in the atmosphere and emptied of their water. In that condition they slide out to the horse latitudes, north and south some hundreds of miles, and then again descend. Coming down, they get compacted into high pressure systems of surpassing dryness. And as the pressure of this falling air increases, so does its temperature. The consequence is extreme permanent aridity along the two latitudinal bands and a first cause for all the world's major deserts.

The second cause is mountains—long ranges of mountains, sprawling out across the path of prevailing winds. These ranges block the movement of moist air, forcing it to ascend over them like a water-skier taking a jump. In the process, that air is cooled to the point where it releases its water. The mountains get deep snow on their peaks and the land to leeward gets what is left: almost nothing. Such a "rain shadow" of dryness may stretch for hundreds of miles downwind, depending on the height of the range. It's no accident, then, that the Sahara is bordered along its northwestern rim by the Atlas Mountains, that the Taklimakan stares up at the Himalayas, that the Patagonian Desert is overshadowed by the Andes.

Ocean currents out of the polar regions work much the same way, sweeping along the windward coastlines of certain continents and putting a chill into the oncoming weather systems before those systems quite reach the land. Abruptly cooled, the air masses drop their water off the coast and arrive inland with little to offer. For instance, the Benguela Current, curling up from Antarctica to lap the southwestern edge of Africa, steals moisture that might otherwise reach the Namib. The Humboldt Current, running cold up the west coast of South America, keeps the Atacama similarly deprived. The California Current, flowing down from Alaska along the Pacific coast as far south as Baja, does its share to promote all-season baseball in Arizona.

Beyond all these causes of dryness, another important factor is wind, helping to shape desert not only through evaporation but also—and more drastically than in any other type of climate zone—by erosion. Powerful winds blow almost constantly into and across any desert, with heavier cold air charging forward to fill the vacuum as hot light air rises away off the desert floor. Desert mountains tend to increase this gustiness, and in some cases to focus winds through canyons and passes for still more extreme effect. In deserts of southwestern North America, they call the wind *chubasco* if it's a fierce rotary hurricane of a thing, whirling up wet and mean out of the tropics and tearing into the hot southern drylands with velocities up to 100 mph, sometimes delivering more than a year's average rainfall in just an afternoon. More innocent little whirlwinds, localized twisters and dust devils, are known as *tornillos*. The steadiest and driest wind out of northern Africa is known as *sirocco*, from an Italian word with an Arabic precursor; the sirocco is what gives southern Europe a sniff of Saharan desiccation. Besides raking away moisture and making life tough for plants and animals, winds work at dismantling mountains, grinding rock fragments into sand, piling the sand into dunes and moving them off like a herd of sheep. The writer and photographer Uwe George has called desert wind "the greatest sandblasting machine on earth," and there is vivid evidence for that notion in any number of desert formations.

The winds and the flash floods are further abetted, in punishing the terrain, by huge fluctuations in surface temperature. A desert thermometer doesn't just go up, way up; it goes wildly up and down, by day and by night, because the clear skies and the lack of vegetation allow so much of the day's solar energy to radiate away after dark. Easy come, easy go, since there's no insulation to slow the transfer of heat. The temperature of the land surface, furthermore, fluctuates even more radically than the air temperature; a dark stone heated to 175°F in the afternoon may cool to 50°F overnight. The result is a constant process of fragmentation— rocks splitting noisily, as though from sheer exasperation.

It is all so elaborately and neatly interconnected. The dryness of desert regions entails clear skies and a paucity of plants; which together entail fierce surface heat by day, bitter chill by night; which leads to rock fracture, crumbling mountains, and the eventual creation of sand. The thermal convection of air brings strong winds, which exacerbate in their turn the aridity and the erosion; the irregularity of rainfall, acting upon soil not anchored by a continuous carpet of plants, creates arroyos, canyons, badlands, rugged mountains; wind and sand collaborate on the dunes and the sculpted rocks. Add to this a team of small, thirst-proof animals like the kangaroo rat, hardy birds like the poorwill, ingeniously appointed reptiles like the sidewinder, arthropods of all menacing variety, and what you have is a desert—a land of hardship, of durable living creatures but not many, of severe beauty, and in some ineffable way, yes, of cleanliness.

In *Seven Pillars of Wisdom*, Lawrence wrote of the desert-dwelling sort of man who had "embraced with all his soul this nakedness too harsh for volunteers, for the reason, felt but inarticulate, that there he found himself indubitably free. He lost material ties, comforts, all superfluities and other complications to achieve a personal liberty which haunted starvation and death. He saw no virtue in poverty herself: he enjoyed the little vices and luxuries—coffee, fresh water, women—which he could still preserve. In his life he had air and winds, sun and light, open spaces and a great emptiness. There was no human effort, no fecundity in Nature: just the heaven above and the unspotted earth beneath. There unconsciously he came near God."

Lawrence was talking about the Bedouin, but it might apply just as well to mad dogs and Englishmen, including himself. No cool distant tone of the anthropologist in those sentences, but an intimacy that sounds autobiographical (except for the sly comment about women, which didn't suit his own taste in "little vices") and more than a bit nostalgic. *There unconsciously he came near God.* Maybe that's what Lawrence meant with his notion of cleanliness: For him, life in the desert had been next to godliness.

Jeremy Bentham, the *Pietà*, and a Precious Few Grayling

RUMOR HAD IT THEY WERE GONE, or nearly gone, killed off in large numbers by dewatering and high temperatures during the bad drought of 1977. The last sizable population of *Thymallus arcticus*, the Arctic grayling, indigenous to a river in the lower forty-eight states: *ppffft*. George Liknes, a graduate student in fisheries biology at Montana State University, was trying to do his master's degree on these besieged grayling of the upper Big Hole River in western Montana, and word passed that his collecting nets, in the late summer of 1978, were coming up empty. The grayling were not where they had been, or if they were, Liknes for some reason wasn't finding them. None at all? "Well," said one worried wildlife biologist, "precious few."

Grayling are not suited for solitude. Like the late lamented passenger pigeon, they are by nature and necessity gregarious, thriving best in rather crowded communities of their own kind. When the size of a population sinks below a certain threshold, grayling are liable to disappear altogether, evidently incapable of successful pairing and reproduction without the advantages supplied by dense aggregation. This may have been what happened in Michigan. Native grayling were extinguished there, rather abruptly, during the 1930s.

The Michigan grayling and the Montana strain had been isolated from each other and from all other grayling for thousands of years. They were glacial relicts, meaning that they had gradually fled southward into open water during the last great freeze of the Pleistocene epoch; then, when the mile-thick flow of ice stopped just this side of the Canadian border and began melting back northward, they were left behind in Michigan and Montana as two separate populations of grayling. These two populations were trapped, as it turned out, cut off by hundreds of miles from what became the primary range of the species, across northern Canada and Alaska. They were stuck in warmish southern habitats occupied more comfortably by competitor species such as cutthroat trout, brook trout, and mountain whitefish, and overlapping the future range of dominance of another problematic species, *Homo sapiens*. Their own future, consequently, was insecure.

The Michigan grayling went first. They had been abundant in the upper part of Michigan's Lower Peninsula and in the Otter River of the Upper Peninsula. One report tells of four people catching three thousand grayling in fourteen days from the Manistee River and hauling most of that catch off to Chicago. By 1935, not surprisingly, the Manistee was barren of grayling. Before long, so was the rest of the state. Saw logs had been floated down rivers at spawning time, stream banks had been stripped of vegetation (causing water temperatures to rise), exotic competing fish (such as brown trout and rainbows) had been introduced, and greedy pressure like that on the Manistee had continued. By 1940 the people of Michigan had just the grayling they were asking for: none.

In Montana, where things tend to happen more slowly, some remnant of the original grayling population has endured, against similar adversities in less intense form, by way of a tenuous balance of losses and gains. Although they have disappeared during the past eighty years from parts of their Montana range, they have meanwhile expanded into some new habitat. More accurately, they have been *introduced* to new habitat, by way of hatchery rearing and planting—the ecological equivalent of forced school busing. As early as 1903, soon after the founding of the Fish Cultural Development Station in Bozeman, the state of Montana got into the business of grayling aquaculture; and for almost sixty years thereafter the planting of hatchery grayling was in great vogue.

The indigenous range of the Montana grayling was in the head-waters of the Missouri River above Great Falls. They were well established in several branches of that grand drainage: the Smith River, the Sun River, the Madison, the Gallatin, and the Jefferson River and its tributaries—notably the Big Hole River. They had evolved mainly as a stream-dwelling species and existed in only a very few Montana lakes. However, they happened to be rather tol-erant of low dissolved-oxygen levels, at least when those levels occurred in cold winter conditions (though not in summer condi-tions, when oxygen was driven out of solution by warming). This made them suitable for stocking in high lakes, where they could get through the winter on what minimal oxygen remained under the ice. In 1909, 50,000 grayling from the Bozeman hatchery were planted in Georgetown Lake. Just a dozen years later, 28 million grayling eggs were collected from Georgetown, to supply hatchery brood for planting elsewhere. And the planting continued: Ennis Lake, Rogers Lake, Mussigbrod Lake, Grebe Lake in Yellowstone National Park. Between 1928 and 1977, millions more grayling were dumped into Georgetown Lake.

Unfortunately, that wasn't all. Back in 1909, hatchery grayling were also planted in the Bitterroot and Flathead Rivers, on the west side of the Continent Divide, in stream waters they had never colonized naturally. It was an innocent experiment, and without large consequences, since the grayling introduced there evidently did not take hold. But then, in what may have seemed a logical extension of all this hatchery rearing and planting, the Big Hole River received a dump of hatchery grayling. The fact that the Big Hole already had a healthy, reproducing population of grayling was not judged to be reason against adding more. From 1937 until 1962, according to the records of the Montana Department of Fish, Wildlife, and Parks (FWP), roughly 5 million grayling from the Anaconda hatchery were poured into the Big Hole, from the town of Divide upstream to the headwaters: hothouse grayling raining down on wild grayling.

This was before FWP biologists had come to the belated realiza-tion that massive planting of hatchery fish in habitat where the same species exists as a reproducing population is the best of all ways to make life miserable for the wild fish. Things are done differently these days, but the mistake was irreversible. The ambitious sequence of plantings was very likely the most disastrous single thing that ever happened to the indigenous grayling of the Big Hole.

At best, each planting instantaneously created tenement conditions of habitat and famine conditions of food supply. In each place where the hatchery truck stopped, the river became a grayling ghetto. At worst, if any of the planted fish survived long enough to breed with one another and interbreed with the wild fish, the whole planting program may have served to degrade the gene pool of the Big Hole grayling, making them less well adapted to the river's particular conditions, less capable of surviving the natural adversities—drought, flood, temperature fluctuation, predation—of their natural habitat.

Then again, it's unlikely that more than a few of those planted grayling did survive long enough to breed. The mortality rate on hatchery grayling planted in rivers is close to 100 percent during the first year, and most don't last even three months, whether or not they are caught by a fisherman. Those planted grayling come, after all, from a small sample of lake-dwelling parents, a sample comprising little genetic variation or inherited capacity for coping with moving water. Reared in the Orwellian circumstances of the hatchery, cooped in concrete troughs, without a beaver or a merganser to harry them, eating Purina trout chow from the hand of man, what chance have they finally in the most challenging of habitats, a mountain river? The term "fish planting" itself is a gross misnomer when applied to dropping grayling or trout into rivers; there is no illusion, even among hatchery people, that many of these plants will ever take root. More realistically, it's like providing an Easter egg hunt for tourists with fishing rods.

In 1962 the Big Hole planting ceased and the remaining wild grayling, those that hadn't died during the famine and tenement periods, were left to get on as best they could. Then came the 1977 drought and, a year later, the George Liknes study. One of Liknes's study sections on the Big Hole was a two-mile stretch downstream from the town of Wisdom to just above the Squaw Creek bridge.

On a certain remote part of the stretch, a rancher had sunk a string of old car bodies to hold his hayfield in place against bank erosion. From that two-mile stretch, using electroshocking collection equipment that is generally reliable for fish censusing, Liknes did not take a single grayling. This came as worrisome news to me, because on a morning in late summer 1975, standing waist-deep within sight of the same string of car bodies and offering no great demonstration of angling skill, I had caught and released thirty-one grayling in four hours. Now they were either gone or in hiding.

Grayling belong to the salmonid family, as cousins of trout and salmon and whitefish. In many ways they seem to be an intermediate form between whitefish and trout, sharing some resemblances with each of those clusters of species. In other ways, they depart uniquely from the salmonid pattern.

The first thing usually noted about them, their identifying character, is the large and beautiful dorsal fin. It sweeps backward twice the length of a trout's, fanning out finally into a trailing lobe, and it is, under certain circumstances, the most exquisitely colorful bit of living matter to be found in the state of Montana: spackled with rows of bright turquoise spots that blend variously to aquamarine and reddish orange toward the front of the fin, a deep hazy shading of iridescent mauve overall, and along the upper edge, in some individuals, a streak of shocking rose. That's how it looks in the wild, or even when the fish is stuck on a hook several inches underwater. Lift the fish into air and the exquisiteness disappears. The bright spots and iridescence drain away at once, the dorsal fin folds down to nothing, and you are holding a drab gunmetal creature that looks very much like a whitefish. The grayling magic vanishes, like a dreamed sibyl, when you pull it to you.

Apart from this dorsal fin, in its optimal display condition, the grayling does resemble that most maligned and misunderstood of Montana creatures, the mountain whitefish, *Prosopium williamsoni*. Both are upholstered, unlike the various trout species, with large, stiff scales—scales you wouldn't want to eat. Both have dull-colored bodies, grayish silver in the grayling, brownish silver in the whitefish, though the grayling does carry as additional adornment a smattering of purplish black spots along its forward flank, playing dimly off the themes in the dorsal fin. Grayling and whitefish are distinguishable (from each other, and from their common salmonid relatives) by the shape of their mouths. A trout has a wide, sweeping, toothy grin. A whitefish mouth is narrow, virtually toothless, and set in a snout that is cartilaginous and pointed, almost like a rat's, which probably contributes to the unpopularity of whitefish among fly fishermen, who don't enjoy disengaging their delicate flies from such rubbery muzzles. The grayling mouth, as you can see if you look closely, is an uneasy compromise between those other two forms: a prim orifice, neither wide nor narrow, set with numerous tiny teeth and fendered with large cartilaginous maxillaries, too short and inoffensive to be fairly called a snout. My point is this: The grayling is one of America's most beautiful fish, but only a few subtle anatomical differences separate it from one of the most ugly. A lesson about pride, I suppose.

But a superfluous lesson, since the grayling by character is anything but overweening. It is dainty and fragile and relatively submissive. With tiny teeth and little moxie, it fails in competition against trout, at least along the southern periphery of its range—and that's another reason for its decline in the Big Hole, where rainbow and brown and brook trout, none of them indigenous, now bully it mercilessly. Like many beautiful creatures that have known fleeting success, the grayling is dumb. It seeks security in gregariousness and these days is liable to find, instead, carnage. When insect food is on the water and the fish are attuned to that fact, a fisherman can stand in one spot, literally without moving his feet, and catch a dozen grayling. Trout are not so foolish. Drag one from a hole and the word will be out to the others. The grayling cannot take such a hint. In the matter of food it is an unshakeable optimist; the distinction between a mayfly on the water's surface and a hook decorated with feathers and floss is lost on it. But this rashness, in the Big Hole for example, might again be partly a consequence (as well as a cause) of its beleaguered circumstances. The exotic trouts, being dominant, seize the choice territorial positions of habitat, and the grayling, pushed off into marginal water where a fish can only with difficulty make a living, may be forced to feed much more recklessly than it otherwise would.

At certain moments the grayling seems even a bit stoic, as though it had seen its own future and made adjustments. This is noticeable from the point of view of the fisherman. A rainbow trout with a hook jerked snug in its mouth will leap as though it is angry, furious—leap maybe five or six times, thrashing the air convulsively each time. If large, it will run upstream, finally to go to the bottom and begin scrabbling its head in the rubble to scrape out the hook. A whitefish, unimaginative and implacable, will usually not jump, will never run, will stay near the bottom and resist with pure loutish muscle. A grayling will jump once, if at all, and remain limp in the air, leaping the way a Victorian matron would swoon into someone's arms—with demure, trusting abandon. Then, possibly after a polite tussle, the grayling will let its head be pulled above the water's surface, turn passively onto its side, and allow itself to be hauled in. Once beaten, a rainbow trout can be coaxed with certain tricks of handling to give you three seconds of docility while you extract the hook so as to release it. A whitefish will struggle like a hysterical pig no matter what. A grayling will simply lie in your hand, pliant and fatalistic, placing itself at your mercy.

So no one has much use for the grayling, not even fishermen. It grows slowly, never as large as a lunker trout, and gives unsatisfactory battle. It is scaly, bony, and not especially good to eat. Montana's fishing regulations allow you to kill five of them from the Big Hole in a day* and five more every day all summer—but what will you do with them? Last year a Butte man returned from a weekend on the river and offered a friend of mine ten grayling to feed his cat. The man had killed them because he caught them, very simple logic, but then realized he had no use for them. This year my friend's cat is dead, through no fault of the grayling, so even that outlet is gone. A grayling does not cook up well, it does not fight well. It happens to have an extravagant dorsal fin, but no one knows why. If you kill one to hang on your wall, its colors will wilt away dishearteningly, and the taxidermist will hand you back a

---

* Or they did at the time this essay first appeared; since then, the regulations have been changed in the grayling's favor.

whitefish in rouge and eye shadow. The grayling, face it, is useless. Like the auk, like the zebra swallowtail, like Angkor Wat.

In June 1978, the U.S. Supreme Court ruled that completion of the Tellico Dam on the Little Tennessee River was prohibited by law, namely the 1973 Endangered Species Act, because the dam would destroy the only known habitat of the snail darter, *Percina tanasi,* a small species of fish belonging to the perch family. One argument in support of this prohibition, perhaps the crucial argument, was that the snail darter's genes might at some time in the future prove useful, even invaluable, to the balance of life on Earth, or at least to the welfare of humanity. If the *Penicillium* fungus had gone extinct when the dodo did, according to this argument, many thousands of additional human beings by now would have died of diphtheria and pneumonia. You could never foresee what you might need, what might prove useful in the line of genetic options, so nothing at all should be squandered, nothing dismissed, nothing relinquished. Thus it was reasoned on behalf of snail darter preservation. The logic is as solid as it is pernicious.

The whole argument by utility may be one of the most dangerous strategic errors that the environmental movement has made. The best reason for saving the snail darter was this: precisely because it is flat useless. That's what makes it special. It wasn't put there, in the Little Tennessee River; it has no ironclad reason for being there; it is simply there. A hydroelectric dam, which can be built in a mere ten years for a mere $119 million, will have utility on its side of the balance against snail darter genes, if not now, then at some future time when the cost of electricity has risen above the cost of recreating (or approximating) the snail darter through genetic engineering. A snail darter arrived at the hard way, the Darwinian way, through millions of years of random variation and natural selection, reaching its culmination in a small homely animal roughly resembling a sculpin, is something far more precious than a net asset in potential utility. What then, exactly? That isn't easy to say, without gibbering in transcendental tones. But something more than a floppy disk storing coded genetic lingo for a rainy day.

Another example: On a Sunday in May 1972, an addled Hungarian named Laszlo Toth jumped a railing in St. Peter's Basilica and took a hammer to Michelangelo's *Pietà*, knocking the nose off the figure of Mary, and part of her lowered eyelid, and her right arm at the elbow. The world groaned. Italian officials charged Toth with crimes worth a maximum total of nine years' imprisonment. Some people, but no one of liberal disposition, declared that capital punishment would be more appropriate. In fact, what probably should have been done was to let Italian police sergeants take Toth into a Roman alley and smack his nose off, and part of his eyelid, and his arm at the elbow, with a hammer. The *Pietà* was at that time 473 years old, the only signed sculpture by the greatest sculptor in human history. I don't know whether Laszlo Toth served the full nine years, but very likely not. Deoclecio Redig de Campos, of the Vatican art-restoration laboratories, said at the time that restoring the sculpture, with glue and stucco and substitute bits of marble, would be "an awesome task that might take three years," but later he cheered up some and amended that to "a matter of months." You and I know better. The Michelangelo *Pietà* is gone. The Michelangelo/de Campos *Pietà* is the one now back on display. There is a large difference. What exactly is the difference? Again hard to say, but it has much to do with the snail darter.

Sage editorialists wrote that Toth's vandalism was viewed by some as an act of leftist political symbolism: "Esthetics must bow to social change, even if in the process the beautiful must be destroyed, as in Paris during *les événements*, when students scrawled across paintings 'No More Masterpieces.' So long as human beings do not eat, we must break up ecclesiastical plate and buy bread." The balance of utility had tipped. The only directly useful form of art, after all, is that which we call pornography.

Still another example: In May 1945 the Target Committee of scientists and ordnance experts from the Manhattan Project met to hash out a list of the best potential Japanese targets for the American atomic bomb. At the top of the list they placed Kyoto, an industrial center inhabited by one million people, which happened also to be the ancient capital of Japan, for eleven centuries the source of much that was beautiful in Japanese civilization, and the site of many gorgeous and sacred Shinto shrines. The target list circulated to a small circle of Washington policy-makers, among whom was Henry L. Stimson, Harry Truman's inherited secretary of war. Stimson was no softie. He was a stubbornly humane old man who had years earlier served as secretary of war under William Howard Taft, then as secretary of state under Herbert Hoover. The notion of targeting Kyoto put his back up. "This is one time I'm going to be the final deciding authority. Nobody's going to tell me what to do on this. On this matter I am the kingpin." And he struck the city of shrines off the list. Truman concurred. Think what you will of the subsequent bombing of Hiroshima—unspeakably barbarous act, most justifiable act under the circumstances, possibly both; still, the sparing of Kyoto, acknowledged as a superior target in military terms, was possibly the most imaginative decision that Harry Truman was ever advised and persuaded to make. In May 1945, the shrines of Kyoto did not enjoy the balance of utility.

"By utility is meant that property in any object, whereby it tends to produce benefit, advantage, pleasure, good, or happiness (all this in the present case comes to the same thing), or (what comes again to the same thing) to prevent the happening of mischief, pain, evil, or unhappiness to the party whose interest is considered: if that party be the community in general, then the happiness of the community; if a particular individual, then the happiness of that individual." This was written by Jeremy Bentham, the English legal scholar of the eighteenth century who founded that school of philosophy known as utilitarianism. He also wrote, in *Principles of Morals and Legislation*, that "an action then may be said to be conformable to the principle of utility . . . when the tendency it has to augment the happiness of the community is greater than any it has to diminish it." In more familiar words, moral tenets and legislation should always be such as to achieve the greatest good for the greatest number. And "the greatest number" has generally been taken to mean (though Bentham himself might not have agreed: see "Animal Rights and Beyond," above) the greatest number of humans.

This is a nefariously sensible philosophy. If it had been adhered to strictly throughout the world since Bentham enunciated it, there would now be no ecclesiastical plate or jeweled papal chalices, no symphony orchestras, no ballet companies, no Picassos, no Apollo moon landings, no well-preserved Kyoto. Had it been retroactive, there would be no Egyptian pyramids, no Taj Mahal, no texts of Plato; nor would there have been any amassing of wealth by Florentine oligarchs and hence no Italian Renaissance; finally, therefore, no *Pietà*, not even a mangled one. And if Bentham's principle of utility—in its economic formulation, or in thermodynamic terms, or even in biomedical ones—is applied today and tomorrow as the ultimate standard for matters of legislation, let alone morals, then there will eventually be no parasitic microbes and no mosquitoes and no man-eating crocodiles and no snail darters and no . . .

But we were talking about the Big Hole grayling. George Liknes was finding few, and none at all near the string of car bodies, and this worried me. I had strong personal feelings toward the grayling of the Big Hole. What sort of feelings? "Proprietary" is not the right word—too presumptuous; rather, something in the vein of "cherishing" and "reliance." I had come to count on the fact, for cheer and solace in a very slight way, that they were there, that they existed—beautiful, dumb, and useless—in the upper reaches of that particular river. It had happened because I had gone up there each year for a number of years—usually in late August, which is the start of autumn in the Big Hole Valley, or in early September—with two hulking Irishmen, brothers. Each year, stealing two days for this pilgrimage just as the first cottonwoods were taking on patches of yellow, we three together visited the grayling.

At that time of year the Big Hole grayling are feeding, mainly in the mornings, on a plague of tiny dark mayflies belonging to the genus *Tricorythodes* and known casually by the shorthand "trikes." A trike is roughly the size of a caraway seed, black-bodied with pale milky wings. Inconsiderable as individuals, they appear on the water by the millions, and the grayling line up (in certain areas) to sip at them. The trike hatch happens every August and September, beginning each morning when the sun begins warming the water, continuing daily for more than a month, and it is one of the reasons thirty-one grayling can be caught in a few hours. The trike hatch was built into my understanding with the Irishmen, an integral part of the yearly ritual. Trike time, time to visit the Big Hole grayling.

Not stalk, not confront, certainly not kill and eat; visit. No great angling thrills attach to catching grayling. You don't fish at them for the satisfaction of fooling a crafty animal on its own terms or fighting a wild little teakettle battle across the tenuous connection of a fine monofilament leader, as you do with trout. The whole context of expectations and rewards is different. You catch grayling to visit them: to hold one carefully in the water, hook freed, dorsal flaring, and gape at the colors, and then watch as it dashes away. This is good for a person, though it could never be the greatest good for the greatest number. I had visited them regularly at trike time with the two Irishmen, including the autumn of the younger brother's divorce, and during the days just before the birth of the older brother's first daughter, and through some personal weather of my own. So I did not want to hear about a Big Hole River that was empty of grayling.

A fair question to the Montana Department of Fish, Wildlife, and Parks is this: If these fish constitute a unique and historic population, a wonderful zoological rarity within the lower forty-eight states, why let a person kill five in a day for cat food? FWP biologists have offered three standard answers: 1) They, the departmental biologists, possessed no reliable data (until George Liknes finished his master's thesis) on the Big Hole grayling, and they do not like to recommend changes in management regulations except on the basis of data; 2) grayling are very fecund—a female will sometimes lay more than 10,000 eggs—and so it's the availability of habitat and infant mortality and competition with trout that limit grayling population levels, not fishing pressure; finally 3) these grayling are glacial relicts, meaning that they have been left behind in this marginal habitat and are naturally doomed to elimination by climate change, with all adverse actions of mankind only accelerating that inevitability.

And yet 1) over a period of twenty-five years, evidently without the basic data that would have revealed such efforts as counterproductive, FWP spent large sums of money to burden the Big Hole grayling with 5 million hatchery outsiders; 2) though fishermen are admittedly not the limiting factor on the total number of grayling in the river, they can easily affect the number of large, successful, genetically gifted spawning stock in the population, since those are precisely the individual fish that fishermen, unlike high temperatures or low oxygen concentrations or competitive trout, kill in disproportionate numbers. There might be money for more vigorous pursuit of data, there might be support for protecting the grayling from cats, but the critical constituency involved here is fishermen, and the balance of utility is not on the side of the grayling. As for 3), not only have the rivers of Montana grown warmer with the end of the Pleistocene, but Earth generally is warming, thanks to human actions and probably also larger geophysical considerations; in fact, our little planet is falling slowly, inexorably out of orbit and into the sun; and the sun itself is meanwhile dying. So all earthly wildlife is doomed to eventual elimination, the world will end, the solar system will end, and mankind is only et cetera.

The year before last, the Irishmen and I missed our visit. The older brother had a second daughter coming, and the younger brother was in Germany, in the army, soon to have a second wife. I could have gone alone but I didn't. So all I knew of *Thymallus arcticus* on the upper Big Hole was what I heard from George Liknes: not good. Through the winter I asked FWP biologists for news of the Big Hole grayling: not good.

Then one day in late August last year, I sneaked away and drove up the Big Hole toward the town of Wisdom, specifically for a visit. I stopped when I saw a promising stretch of water, a spot I had never fished or even noticed before, though it wasn't too far from the string of car bodies. I didn't know what I would find, if anything. On the third cast I made contact with a twelve-inch grayling, largish for the Big Hole within my memory. Between sun-on-the-water and noon, using a small fly resembling a *Tricorythodes*, I caught and released as many grayling as ever. As many as I needed.

I could tell you where to look for them. I could suggest how you might fish for them, but that's not the point here. You can find them yourself if you need to. Likewise, it's tempting to suggest where you might send letters, whom you might pester, what pressures you might apply on behalf of these useless fish; again, not exactly the point. I merely wanted to let you know: They are there.

Irishmen, the grayling are still there, yes. Please listen, the rest of you: They are there, the Big Hole grayling. At least for now.

Yin and Yang in the Tularosa Basin

IN THE OUTBACK OF SOUTHERN NEW MEXICO, laid down across a thousand square miles of otherwise unexceptional desert, there is a message.

It is gigantic and stark. A simple design, vaguely familiar, executed in unearthly black-and-white against the brown desert ground. You could see it from the moon with a cheap telescope. Nevertheless, it is more cryptic than Stonehenge. An army of zealous Chinese masons might have spent their lifetimes erecting such a thing—but no, that's not where it came from. The full design includes four elements, only two of those manmade, the others attributable to natural processes of the sort that are loosely called "acts of God." In the desert, as Moses found out, God seems to act with an especially bold hand. The constituent materials in this particular case are basalt rock and gypsum sand; black and white, rough and smooth, hard and soft; dissimilar as fire and ice. The message is drawn in high contrast. The text is clear but the meaning is not.

A dune field of startling whiteness, called the White Sands, sprawls out in giant amoeboid shape, creeping northeastward with the winds. A hardened black flow of recent volcanic lava, called the Carrizozo Malpais, stretches southwestward down the gentle incline along which gravity pulled it from its point of emergence through the earthly crust. Near an old ranch site known as Three

Rivers, the leading edge of the whiteness approaches the forward lip of the blackness, leaving a gap of not many miles. Through that gap runs a narrow range road, open for public travel only one day a year. Where the road goes, and why armed men in guardhouses monitor its disuse, are questions for later. The oddities about this place will have to be taken in turn. We are in the Tularosa Basin, a sunken valley full of saltbush and lizards and history, gypsum and lava, plus more than its share of preternatural romance, lying half-way between Las Cruces and Roswell on the way to nowhere at all. We are here, first of all, for the big design.

The design: Superimposed on the desert by a convergence of geological accidents, it is an unmistakable yin and yang, a huge magnification of the Taoist emblem that stands for the paradox of dialectical oneness—two teardrops bound complementarily into a circle, dark and light, head to tail, representing the unity within which all worldly flux remains balanced. In this particular case, the emblem is as large as Long Island. From the moon or beyond, with your telescope, you might take it for a symbol of harmony. The confusion would be understandable.

An abundance of gypsum was the earliest of those geological accidents.

Gypsum is curious stuff from which to make a dune field. In scientific notation it is $CaSO_4 \cdot 2H_2O$, meaning simply the mineral calcium sulphate, bound up in crystalline form with a proportion of plain water. More familiarly, it is the main ingredient of plaster of paris. Under ideal conditions, falling out of a heavy solution, it grows into elegant daggerlike crystals called selenite, which are more or less clear or amber, depending on purity. But as erosional forces break selenite down into small granules—sand—the faces of those granules, being relatively soft, become scratched. The scratches scatter light. The result is whiteness. You hear the name White Sands, but until you make your own pilgrimage, until you lose yourself in the heart of these dunes with only a canteen and a compass, the words are unlikely to register as they should.

Take them literally. White sands. Whiteness like ivory. Like the sun-bleached skull of a lost desert cow. Whiteness like January in the Absaroka Mountains between Montana and Wyoming, 200 yards above timberline. Actually there is nothing and nowhere else quite like White Sands, the world's largest expanse of wind-blown gypsum. Nowhere else on Earth where you can surround yourself with such profound whiteness and still be in danger of snakebite. The white dunes began forming perhaps 25,000 years ago, but the gypsum has been here much longer.

It was deposited during the late Paleozoic era, gypsum-rich layers of sedimentary rock left behind from gypsum-rich seawaters as the long cycles of climate moved an ocean coast back and forth over what is now southern New Mexico. Other sediments were left in the course of other cycles, burying the gypsum beneath hundreds of feet of limestone and shale. It might have stayed there, inert and hidden, like most of the gypsum in Earth's crust, if not for the next accident. Geological pressures that were creating the Rocky Mountains also caused this particular area to buckle upward into a high rounded plateau. Roughly 10 million years ago, another shifting of pressures caused a pair of fault lines to develop, running north-south for a hundred miles; and along these faults, the plateau fell like a startled cake. The parallel fault lines became a matched set of continuous escarpments, mountainous walls, facing each other over a sunken basin. On the east side looking west was what's now called the Sacramento range; on the west looking east, the San Andres. In between was the Tularosa.

To the north and south also, the Tularosa was blocked by high ground. Like many valleys in the desert country of the West, it had no outlet to the sea. So when the next cycle of wetness began, this basin turned into a vast lake.

Erosional torrents flowed down from the mountains to fill it, and (because the mineral is easily eroded and highly soluble) those waters carried gypsum. The lake in its turn became gypsum-rich. Then our most recent age of relative drought dried it away to almost nothing. The waters shrank gradually back to the lowest spot in the basin, a small area toward the southwest corner, and as the big lake gave up the ghost, it also gave up the gypsum. Meanwhile groundwater flow from the upper end of the basin also carried dissolved gypsum, underground, toward the same spot. These days Lake Lucero is never more than a briny puddle, shin-deep at the end of the monsoon season. For most of the year, it is only a dry bed. But it derives a mute dignity from being the source of the White Sands.

With each cycle of evaporation—nowadays, as for thousands of years—small selenite crystals bloom magically along the puddle's margins. Seasonal windstorms roar out of the southwest, through gaps in the San Andres Mountains, grinding the crystals to sand. And so the dunes gather themselves, rise, and move.

The nimblest of them advance about 30 feet in a year. Others travel more slowly. Today the sand is spread over an area of almost 300 square miles, enough to constitute a distinct ecosystem with its own patterns of organismic association, its (temporarily) stabilized zones of vegetation, its uniquely adapted races of animal. A whole world of life hides at the heart of the White Sands. But despite 25,000 years of shared history, there is a gypsy quality to that life. The restlessness of the dunes imposes special demands. The very ground here is in motion. *We are the dunes: we cover all. You must move along with us, or get out of the way, or die.* The animals, even the plants, manage to cope with that imperative in their own patient, mobile ways.

The White Sands are gliding northeastward, inexorably, toward that shape of blackness in the near distance.

The entire Tularosa Basin is tilted slightly downhill toward the southwest, like a great earthen flume. Elevations above sea level vary, from around 6,000 feet at the northern end, to 5,400 near the town of Carrizozo, down to 4,000 feet at Lake Lucero. That incline is the simplest—and least mystical—explanation for what the lava did. Vomiting suddenly up from underground one stormy day, molten black rock flowed steaming and hissing off toward exactly that point from which came the White Sands, as though some dark subterranean animus of alarming proportion were seeking to reunite itself, or maybe do battle, with its antipodal twin.

Halfway there, the lava grew cool and viscid. Wrinkled with corrugations, pocked with gas bubbles, it slowed to a sloppy halt, congealing like a runnel of candle wax. It had traveled 44 miles on a line and lapped out across 120 square miles of desert. In some places it was 100 feet deep.

The source of all this lava was a volcanic vent near the north end of the valley, a spigot-hole down to the planet's liquid innards. The site of the vent is still marked, above the rest of the lava field, by a high cone of cinder known on the maps as Little Black Peak. Probably the lava poured out of this hole in two separate episodes, closely spaced. How long ago? The geologists can only guess. Maybe 2,000 years. Maybe less. Without question, the Carrizozo Malpais (a Spanish word for "badlands") is one of the youngest and best preserved lava fields in the continental United States.

Its most striking characteristic is texture. The whole process of liquid rock flowing over rough terrain—cooling differentially from the outside in, piling up on itself into ropy corrugations and eddies, trapping gas bubbles under thin-lidded domes—is captured as in a snapshot. Many of the big bubbles have collapsed to chasms, tiger pits 30 feet deep. Fissures have appeared. At some places the basalt, light and brittle stuff, has been broken into shards by the crowbar of weather. But in the full scope of geologic time, weathering has scarcely begun. The crowbar has been applied but not yet the grinder, still less the emery cloth. This formation is jagged and raucous and therefore, we know deductively, very new.

Some experts date the eruption to around A.D. 500. Archeological evidence adds another interesting angle: Whenever the thing happened, apparently humans were already in the valley to witness it.

They seem to have been a pueblo-building people, sedentary agriculturalists, probably members of what now is referred to as the Mogollon culture. Eventually they disappeared, or were driven out, to be replaced by ancestors of the Mescalero Apaches, a very different bunch. No one knows just why the Mogollon folk went away. Possibly their departure was related to gradual changes in climate and water supply that were incompatible with their farming practices. Or it might have been war. Or something else.

One early commentator, writing in *Science* back in 1885, offered this: "A stream of no mean size seems to have once run down this valley. Not only has it now disappeared, but its bed is covered by lava and loose soil sometimes to great depths. As to the cause of the disappearance, it may have some connection with the tradition of the Indians which tells of a year of fire, when this valley was so filled with flames and poisonous gases as to be made uninhabitable."

In 1966 the state of New Mexico set aside a small tract of the lava flow for public enjoyment and edification. A parking lot and a set of restrooms have been added, also a short loop trail out through the Malpais, complete with number-keyed features of geological and botanical interest for which commentary is supplied in a printed brochure. The place is fascinating, and largely unappreciated. To most people who visit it—and there aren't many—it is just a rest stop on the godforsaken two-lane between Carrizozo and another sleepy town. It is called, quite aptly, Valley of Fires State Park.

Back down at White Sands, public enjoyment and edification are overseen by the U.S. Interior Department. Early in 1933 (it seems to have been one of Herbert Hoover's last official acts) a presidential proclamation was signed, establishing White Sands National Monument. The borders of the monument encompass a sizable swatch of the dune field, though by no means all of it. Lake Lucero is included, and enough area to provide a good sampling of the different dune types and the biotic communities that exist among them, but the leading edge of the dune field is far outside the monument's northern boundary. At large, beyond Park Service jurisdiction; off on its own reckless chase.

And surrounding these two modest administrative units—literally arching over them through the sky—is another official fiefdom, one whose appointed mission does not include edifying the public, except perhaps in the most indirect way: White Sands Missile Range. Run by the army, it is America's largest land-area shooting gallery for the testing of aeronautical and ballistic weaponry. Forty miles wide, stretching north and south for a hundred, it overlaps the whole Tularosa Basin almost exactly. Someone decided, back in 1945, that this is what the Tularosa was good for. *Hey, let's use that big piece of empty desert on the east side of the San Andres Mountains. It's perfect. Who will ever miss it?* Virtually no one has. Outsiders don't often come here to do their communing with God or nature. And the residents of Alamogordo, the largest town in the basin, seem generally to welcome the military dollars.

Today high-altitude research rockets go up and come down over the missile range. Cruise missiles under development by the navy and the air force are launched from bombers, finding their way with magical sentience to targets among the creosote bushes. Drone aircraft—large jet-powered skeet, guided remotely—are blasted out of the sky by the latest and best in air-defense ordnance, leaving small scraps of their debris to flutter down onto the white dunes like titanium confetti. A laser-guided artillery projectile is fired off toward another corner of desert. From Fort Bliss, down near El Paso, Pershing II missiles fly up here on their intermediate-range trajectories, sailing over White Sands as though it were Poland, heading on toward the Malpais. Ground-to-ground missiles, air-to-air, air-to-ground and vice versa, every combination a country could need or believe it needs, the names themselves resonant with mythology and mystery and stouthearted martial precision: Nike, Talos, Tomahawk, Lance, Copperhead, Patriot, Stinger.

From one point of view, it is not so doleful as it might seem. The White Sands is a delicate ecosystem where the margin of survival is narrow, for both animals and plants, and disturbances are not easily repaired. The entire chain of life there depends on a very slow process of soil formation and preliminary vegetal growth that occurs only in flat lowland areas between the active dunes. The slightest vehicle traffic (even heavy foot traffic) leaves scars across those flats that remain visible for decades, and an abusive degree of traffic could pull the bottom out from under the whole biotic community. Park Service regulations prohibit off-road vehicular traffic within the monument, but that only covers a fraction of the dune field. Which is why Dr. William Reid, an ecologist from the University of Texas at El Paso who knows the White Sands as well as anyone, says: "Some people resent the missile range. I think it's a great boon in disguise. It's what saves the dunes and the interdune areas from the people in four-by-fours." A fragile environment gets army protection from demented Sunday dirt-racers, and the occasional flaming rocket crash is, true enough, a small ecological price to pay.

So the innermost precincts of the Tularosa Basin remain unapproachable. You do not wander at will onto the White Sands Missile Range, either by truck or on foot. Fences and brusque signs warn you back. Electronic border sensors notice you. Polite security officers appear, carrying shotguns. The message is *Keep out. This is secret stuff and we're busy. Besides, you could get hurt.* Then, one day each year, the gate on a certain road swings open, and hundreds of vehicles drive through.

This annual motorcade begins from the parking lot of a Kmart on the north edge of the town of Alamogordo, where the cars and the pickups and the RVs with out-of-state plates have gathered and pulled into file by 7:30 A.M. of the appointed day. It is the first Saturday in October, warm and bright, indecently good weather for a picnic or a football game, and though neither of those is the purpose today, still the atmosphere is just a bit festive. A few children have been dragged along for the occasion, by parents or grandparents, and the kids burn off their energy dodging between bumpers, just as they would amid any dull gathering of stalled cars, for a fair or a funeral. Some folks have brought hampers of food; those in the open-topped sports cars wear jaunty hats. History is the sole attraction, but most of the people here assembled have come out in order to feel good about the particular moment of history in question. A much smaller segment—and they will be distinguishable when the picket signs appear—have come out in order to feel guilty and worried and bad. Only a few of us have come out, fully premeditated, in order to feel confused and ambivalent.

The men from the Alamogordo Chamber of Commerce (cosponsors of the day's tour, along with the missile range management) do their directing of traffic with brisk, cheerful authority. After a few years' practice, they feel that they have this thing down to a stroll. It is the Alamogordo equivalent of a big pancake breakfast, or a sweet-pea festival, or a rattlesnake roundup: an expression, and a reinforcement, of civic pride. The 250 autos move off right on schedule. The missile range fellows are holding their fire, but only so long. Just six hours have been set aside for these pilgrims to drive out across the Tularosa, deep into missile range property, on a thin asphalt road running between the White Sands and the black lava, to a place called Trinity Site; to see the marker there, to hear the speeches; and then to get themselves back out of the line of fire.

Of course Trinity Site is the patch of woebegone desert from which Alamogordo, by metonymic incrimination, draws its greatest fame. Actually the site is sixty miles northwest of Alamogordo, shielded behind a stark upthrust of rock known as Oscura Peak. Here, on the morning of July 16, 1945, the nuclear age dawned gaudily. Robert Oppenheimer and his coven of young wizard physicists and mathematicians and engineers, from up in Los Alamos, had chosen the spot because of its sheer desolateness as a good place to test what was then still considered a dubious, improbable gadget.

The test was called Trinity. That code name had been supplied by Oppenheimer, from some free-associative inspiration about which he was ever afterward vague. In a letter, years later, he wrote: "Why I chose the name is not clear, but I know what thoughts were in my mind. There is a poem of John Donne, written just before his death, which I know and love. From it a quotation:

> *As West and East*
> *In all flat Maps—and I am one—are one,*
> *So death doth touch the Resurrection.*

That still does not make Trinity; but in another, better known devotional poem Donne opens, 'Batter my heart, three person'd God.' Beyond this, I have no clues whatever." Such an arcane and obfuscatory explanation was, for Robert Oppenheimer, utterly characteristic.

A different Los Alamos scientist was directly in charge of the Trinity operation, but the two senior officials present were Oppenheimer, the laboratory director, and General Leslie Groves. Groves was a portly career soldier from the Army Corps of Engineers, a man with a large ego and an abrasive personality who had been thwarted in his hope of seeing overseas duty during World War II and who wound up instead, to his dismay, in command of the Manhattan Project. Groves's original mandate seems to have been limited, a simple engineering task in the corps tradition: to build the laboratories in which others would build the Bomb. But by degrees, in the wartime confusion, he filled a vacuum to become supreme potentate of the entire effort. It was Groves who had picked Robert Oppenheimer to run Los Alamos, the final-stage lab where the details of the weapon's design were worked out—picked him despite advice from the FBI that Oppenheimer was a security risk. Groves chose to ignore the FBI charges, but not without letting Oppenheimer know he had heard them. Which gave the general a certain leverage. Despite (or perhaps partly because of) that leverage, the two men had settled into a harmonious and effective working relationship.

In character, background, capabilities, these two couldn't have been more dissimilar. Oppenheimer was an intellectual of broad interests and surprisingly disparate eruditions, who read the classics of Greek and Sanskrit and Spanish literature, loved poetry, carefully studied the work of Karl Marx to see for himself what was there. He had come out of Harvard and done graduate and postdoctoral work at some of the best universities of Europe. He was an epicure. During the 1930s he had been active in leftist causes, generous financially and with his time, never quite a card-carrying Communist but sympathetic with much that the party was doing. Leslie Groves was an engineer and a soldier, period. Son of an austere Presbyterian minister who was himself also an army man, a chaplain to the Fourteenth Infantry, Groves grew up on military posts and went to West Point. Like his father, he was something of a martinet. He was bullish and direct-minded and good at pushing straightforward jobs to completion; not so good at dealing with people. Impatient with psychological complexity. He knew nothing at all about nuclear physics when he was picked to ramrod the A-bomb project, but he never allowed that to dampen his confidence in his own authority. Sometimes he was obtuse; sometimes, in the view of certain scientists, he behaved like a boob. Groves's military deputy on the project said later: "He's the biggest sonovabitch I've ever met in my life, but also one of the most capable individuals. . . . I hated his guts and so did everyone else but we had our form of understanding." Groves was a large man, well upholstered in flesh. Robert Oppenheimer was gangly and emaciated. Both of them could be arrogant, both could be quick to judge. By objective criteria, they should have been expected to loathe and distrust each other wholeheartedly. But it didn't unfold that way. As an unlikely partnership, the physicist and the soldier also evidently had their own form of understanding.

And so Trinity happened, a great success.

Originally the test firing was scheduled for four o'clock on the morning of July 16, a Monday. By Saturday afternoon the bomb had been assembled on the site—its plutonium core inserted delicately into the larger casing—and hoisted up to the top of its 100-foot girder tower. Electrical detonators were connected at sixty-four points along the outside of the metal sphere, small taps plugged onto the surface, a tangle of crisscrossing wires, as though the monstrous and inscrutable thing were having its mind read by electroencephalograph. What was it thinking? What did it know? Notwithstanding the elaborate electronics, that could be found out only the hard way. Before sundown on the last afternoon, Robert Oppenheimer himself climbed the tower, alone, for a last look at this device he had guided into being. About the same time, General Groves arrived at the site. Then the weather turned bad.

Throughout Sunday night the Trinity gadget sat atop its steel tower in the midst of a desert storm, a raucous overture of thunder and lightning and wind-driven rain. No one seems to have gotten an hour's sleep except Leslie Groves. Oppenheimer paced and fretted. One bolt of lightning striking the tower might not have detonated the bomb, but it certainly would have destroyed the electrical circuitry and caused a major delay. Any delay now was dreaded, because Harry Truman at Potsdam was eager for news about this far-fetched atomic weapon, and the test results would tell him how to deal with Stalin concerning the continuing war against Japan. But lightning wasn't the only meteorological problem. There was also concern that storm clouds would carry large doses of fallout onto population centers downwind. Amarillo was 300 miles away, and more immediately, the small town of Carrizozo lay just east across the Tularosa Basin. Groves woke from his nap and consulted with Oppenheimer.

Then around 4 A.M. the rain stopped. The countdown resumed.

A young scientist named Joe McKibben, responsible for the remote electrical signals, threw a switch at minus 45 seconds that locked the whole system into an automatic timer. Though not unstoppable, the event was now progressing on its own robotic momentum, without active human control. In the command bunker and back at the base camp, people lay down on their bellies with their feet toward the tower, a position of backward obeisance. At 5:29:45 A.M. Mountain War Time, the wrapper of high explosives squeezed down on the plutonium core. Neutrons ricocheted, atoms split; a chain reaction ensued.

The report that General Groves sent off at once to Potsdam said: "For a brief period there was a lighting effect within a radius of 20 miles equal to several suns at midday; a huge ball of fire was formed which lasted for several seconds. This ball mushroomed and rose to a height of over 10,000 feet before it dimmed. The light from the explosion was seen clearly at Albuquerque, Santa Fe, Silver City, El Paso, and other points generally to about 180 miles away. . . . A massive cloud was formed which surged and billowed upward with tremendous power, reaching the sub-stratosphere at an elevation of 41,000 feet, 36,000 feet above the ground, in about five minutes, breaking without interruption through a temperature inversion at 17,000 feet which most of the scientists thought would stop it. . . . Huge concentrations of highly radioactive materials resulted from the fission and were contained in this cloud." That account, concrete and dispassionate, reached Truman by courier. In a quiet moment sometime afterward, Robert Oppenheimer offered his own version of the moment: "A few people laughed, a few people cried. Most people were silent." Then he quoted another bit of poetry.

The command bunker and the base camp are long since gone. Oppenheimer and Groves are gone. Hiroshima and Nagasaki are not the same cities they were. But Trinity Site was just a spot out in the desert, and so it remains.

On the first Saturday of October, four decades later, you can still see the frizzled steel stumps left behind when the tower was vaporized. You can still pick up a chirpy reading on a Geiger counter. You can hear speeches by the commander of the missile range and the president of the Chamber of Commerce, and you can join in a prayer with an army chaplain who says: ". . . and guide us, Lord, that we may then begin beating our plowshares into . . . uh, beating our swords into plowshares." The confusion is understandable. You can even chat with Joe McKibben, the man who threw that last switch, now a retirement-age gentleman in casual clothes with a friendly and slight dotty manner, who happens to be back for the tour this year himself. McKibben is genial about answering questions, but there is an unreachable look in his eyes.

He says: "Well, you have to wonder how it would have gone if some things had been different."

There is one more stop on the Tularosa circuit; one more element in the Tularosa design. This one could easily be overlooked, so watch carefully for a small sign along the two-lane that runs north out of Alamogordo toward Carrizozo. Again you will be about halfway from White Sands to the Malpais, but with no motorcade surrounding you now, no men waving you forward with flashlights, no guard stations or crossbars. You stay alert for an inconspicuous junction, at the corner of which sits the bulky white shape of what once was a merry establishment. The building is boarded up; lettering on the window says 3 RIVERS AND DEVIANTS M/C CLUBHOUSE. PRIVATE. BEWARE OF DOG. An M/C, in case you need telling, is a motorcycle club. The 3 Rivers boys and the Deviants (evidently a syncretic group) are not presently in session. But that's your landmark to turn right.

You drive east for another five miles on dirt, into the foothills of the Sacramento Range. You park and begin walking, up the crest of a sharp north-south ridge. Another sign shows you the way. Welcome to the Three Rivers Petroglyph Site, famous among archeologists of the Southwest and unknown to almost everyone else.

A community of the Mogollon people, those pacific agriculturalists, lived here at the start of the present millennium for a span of about four hundred years. Their village was down lower in the Three Rivers drainage; this exposed ridge seems to have served as a lookout, from which they could spot game or approaching enemies far out in the Tularosa Basin. The view is indeed good. Gazing westward across the desert, you can see the great white shape of gypsum and the great black shape of lava. You can see the barren upthrust that is Oscura Peak, and if a fireball 10,000 feet high were suddenly to blossom behind it, you would sure as hell see that too. Mogollon scouts may have spent many hours and weeks and years up on this boulder-toothed ridge, watching. After four centuries, though, the whole community disappeared.

Probably they migrated north, out of the Tularosa. They could have been fleeing a drought. Or their exodus may have been linked to that dolorous remnant of collective memory, the one telling about "a year of fire, when this valley was so filled with flames and poisonous gases as to be made uninhabitable." Maybe they saw something distressful. A sudden thunderous ebullition of cinder and smoke and liquidy black rock, for instance. Or who knows what.

All they left behind were a few potsherds, a few fallen adobe pit-houses, and about five thousand rock carvings, scratched and chipped onto the boulder faces along that ridge.

Some of these carvings are far more artful than others. Some are vividly representational—a bighorn sheep impaled by three arrows—and some aren't. There are a few human figures, but no romantic and elegant portrayals of prowess in battle or hunting; mainly large ovoid heads, wide-eyed and jug-eared. No warriors on horseback. Aside from those arrows in the bighorn, weaponry is conspicuously absent. Animal portraits abound, especially birds and horned mammals and even a few fish. Also there are carven images of tracks: bear paws, bird prints as though in mud, human foot shapes. The preponderance of the petroglyphs on the ridge, though, are abstract designs.

Among these, the most common motif, appearing in many variations, is a circle or several concentric circles surrounded by a ring of dots. Similar circle-and-dot patterns are known from Mogollon petroglyph sites throughout the Southwest, but they seem to have held a special fascination for the artists at Three Rivers. You can see in them almost anything you might choose: a circle of family members, the solar system, the nucleus and electrons of an atom.

Walking the ridge trail up among these carvings, you find another arresting motif. Having made a lucky detour off the main path, watching for rattlesnakes as you step, you notice it first on the western side of a large dark boulder cropping out high on the ridge's westernmost knob. This design is more elaborate and sophisticated than others, even beautiful, and something about it stops you short:

It recurs, in its variations, a dozen times on the ridge. In a few instances it is less squarish, more curvilinear; in several it is done with linked or interlocking spirals. Despite transmutations, in each the essence is unmistakable.

Evidently the people at Three Rivers had a concept of yin and yang. They drew rock pictures of dialectical oneness. They cherished some notion—maybe it was only wishful—of a unity within which all worldly flux remains balanced.

What did that mean to them? Obviously, we don't know. They watched and they drew and then they departed. But the yin-and-yang notion is easily reduced to truism, and these figures, like the dotted circles, can be taken to represent almost anything. John Donne's idea, for instance: "As West and East . . . are one, / So death doth touch the Resurrection." Or another idea, maybe in this case more applicable: "For they have sown the wind, and they shall reap the whirlwind."

Now suddenly a pair of turkey vultures wheel into view above you, cruising on thermals that rise off the west slope of the Three Rivers ridge. With typical lazy grace, they are scouting for a meal. One of the vultures sweeps closer to scrutinize you.

This bird pauses, holding position not 30 feet over your head, like a kite on a short string. It seems unsure whether to take you for a pile of dead meat. And you *are* sitting quite still. The confusion is understandable.

AFTER
THOUGHTS

Planet of Weeds

HOPE IS A DUTY from which paleontologists are exempt. Their job is to take the long view, the cold and stony view, of triumphs and catastrophes in the history of life. They study the fossil record, that erratic selection of petrified shells, carapaces, bones, teeth, tree trunks, leaves, pollen, and other biological relics, and from it they attempt to discern the lost secrets of time, the big patterns of stasis and change, the trends of innovation and adaptation and refinement and decline that have blown like sea winds among ancient creatures in ancient ecosystems. Although life is their subject, death and burial supply all their data. They're the coroners of biology. This gives to paleontologists a certain distance, a hyperopic perspective beyond the reach of anxiety over outcomes of the struggles they chronicle. If hope is the thing with feathers, as Emily Dickinson said, then it's good to remember that feathers don't generally fossilize well. In lieu of hope and despair, paleontologists have a highly developed sense of cyclicity. That's why I recently went to Chicago, with a handful of urgently grim questions, and called on a paleontologist named David Jablonski. I wanted answers unvarnished with obligatory hope.

Jablonski is a big-pattern man, a macroevolutionist, who works fastidiously from the particular to the very broad. He's an expert on the morphology and distribution of marine bivalves and gastropods—or clams and snails, as he calls them when speaking casually. He sifts through the record of these mollusk lineages, preserved in rock and later harvested into museum drawers, to extract ideas about the origin of novelty. His attention roams back through 600 million years of time. His special skill involves framing large, resonant questions that can be answered with small, lithified clamshells. For instance: By what combinations of causal factor and sheer chance have the great evolutionary innovations arisen? How quickly have those innovations taken hold? How long have they abided? He's also interested in extinction, the converse of abidance, the yang to evolution's yin. Why do some species survive for a long time, he wonders, whereas others die out much sooner? And why has the rate of extinction—low throughout most of Earth's history—spiked upward cataclysmically on just a few occasions? How do those cataclysmic episodes, known in the trade as mass extinctions, differ in kind as well as degree from the gradual process of species extinction during the millions of years between? Can what struck in the past strike again?

The concept of mass extinction implies a biological crisis that spanned large parts of the planet and in a relatively short time eradicated a sizable number of species from a variety of groups. There's no absolute threshold of magnitude, and dozens of different episodes in geologic history might qualify, but five big ones stand out: Ordovician, Devonian, Permian, Triassic, Cretaceous. The Ordovician extinction, 439 million years ago, entailed the disappearance of roughly 85 percent of marine animal species—and that was before there *were* any animals on land. The Devonian extinction, 367 million years ago, seems to have been almost as severe. About 245 million years ago came the Permian extinction, the worst ever, claiming 95 percent of all known animal species and therefore almost wiping out the animal kingdom altogether. The Triassic, 208 million years ago, was bad again, though not nearly so bad as the Permian. The most recent was the Cretaceous extinction (sometimes called the K-T event, because it defines the boundary between two geologic periods, with K for Cretaceous, never mind why, and T for Tertiary), familiar even to schoolchildren because it ended the age of dinosaurs. Less familiarly, the K-T event also brought extinction of the marine reptiles and the ammonites, as well as major losses of species among fish, mammals, amphibians, sea urchins, and other groups, totaling 76 percent of all species. In between these five episodes occurred some lesser mass extinctions, and throughout the intervening lulls extinction continued too—but at a much slower pace, known as the background rate, claiming only about one species in any major group every million years. At the background rate, extinction is infrequent enough to be counterbalanced by the evolution of new species. Each of the five major episodes, in contrast, represents a drastic net loss of species diversity, a deep trough of biological impoverishment from which Earth only slowly recovered. How slowly? How long is the time lag between a nadir of impoverishment and a recovery to ecological fullness? That's another of Jablonski's research interests. His rough estimates run to 5 or 10 million years. What drew me to this man's work, and then to his doorstep, was his special competence on mass extinctions and his willingness to discuss the notion that a sixth one is in progress now.

Some people will tell you that we as a species, *Homo sapiens*, all 6 billion of us in our collective impact, are destroying the world. Me, I won't tell you that, because "the world" is so vague, whereas what we are or aren't destroying is quite specific. Some people will tell you that we are rampaging suicidally toward a degree of global wreckage that will result in our own extinction. I won't tell you that either. Some people say that the environment will be the paramount political and social concern of the twenty-first century, but what they mean by "the environment" is anyone's guess. Polluted air? Polluted water? Acid rain? Toxic wastes left to burble beneath neighborhood houses and malls? A frayed skein of ozone over Antarctica? Global warming, driven by the greenhouse gases emitted from smokestacks and cars? None of these concerns is of itself the big one, paleontological in scope, though some (notably, climate change) are closely entangled with it. If the world's air is clean for humans to breathe but supports no birds or butterflies, if the world's waters are pure for humans to drink but contains no fish or crustaceans or diatoms, have we solved our environmental problems? Well, I suppose so, at least as environmentalism is commonly construed. That clumsy, confused, and presumptuous formulation "the environment" implies viewing air, water, soil, forests, rivers, swamps, deserts, and oceans as merely a milieu within which something important is set: human life, human history. But what's at issue in fact is not an environment; it's a living world.

Here instead is what I'd like to tell you: The consensus among conscientious biologists is that we're headed into another mass extinction, a vale of biological impoverishment commensurate with the big five. Many experts remain hopeful that we can brake that descent, but my own view is that we're likely to go all the way down. I visited David Jablonski to ask what we might see at the bottom.

On a hot summer morning, Jablonski is busy in his office on the second floor of the Hinds Geophysical Laboratory at the University of Chicago. It's a large open room furnished in tall bookshelves, tables piled with books, stacks of paper standing knee-high off the floor. The walls are mostly bare, aside from a chart of the geologic time scale, a clipped cartoon of dancing tyrannosaurs in red sneakers, and a poster from a Rodin exhibition, quietly appropriate to the overall theme of eloquent stone. Jablonski is a lean forty-five-year-old man with a dark full beard. Educated at Columbia and Yale, he came to Chicago in 1985 and has helped make its paleontology program perhaps the country's best. Although in not many hours he'll be leaving on a trip to Alaska, he has been cordial about agreeing to this chat. Stepping carefully, we move among the piled journals, reprints, and photocopies. Every pile represents a different research question, he tells me. "I juggle a lot of these things all at once because they feed into one another." That's exactly why I've come: for a little rigorous intellectual synergy.

Let's talk about mass extinctions, I say. When did someone first realize that the concept might apply to current events, not just to the Permian or the Cretaceous?

He begins sorting back through memory. In the early and middle 1980s, he recalls, there occurred a handful of symposiums and lecture series on the subject of extinction, for which the rosters included both paleontologists and what he whimsically calls "neon-tologists," meaning biologists who study creatures that are still alive. An event was held at the Field Museum here in Chicago, with support from the National Science Foundation, that attracted four hundred scientists. Another, at which Jablonski himself spoke, took place at the New England Aquarium, in Boston. "The chronology is a little hazy for me," he says. "But one that I found most impressive was the Elliott meeting, the one in Flagstaff." David K. Elliott, of the geology department at Northern Arizona University, had pulled that gathering together in August of 1983 and later edited the invited papers into a volume titled *Dynamics of Extinction*. The headliner among the neontologists was Paul Ehrlich, eminent as an ecologist, widely famed for his best-selling jeremiad *The Population Bomb*, and coauthor of a 1981 book on human-caused extinctions. Ehrlich spoke mainly about birds, mammals, and butterflies, sketching the severity of the larger problem and offering suggestions about what could be done. For the paleontological perspective, it was Jablonski again and a few others, including John Sepkoski and David Raup, later to be his colleagues at the University of Chicago. Sepkoski and Raup, while sorting through a huge body of data on the life spans of fossil groups, had lately noticed a startlingly regular pattern of recurrence—at about 26-million-year intervals—in the timing of large and medium-sized mass extinctions. The Sepkoski-Raup paper at Flagstaff, building on their own earlier work as well as the hot new idea that an asteroid impact had killed off the dinosaurs, suggested a dramatic hypothesis for explaining those recurrent mass extinctions: that maybe an invisible twin star (an "undetected companion," they called it) orbits mutually around our sun, returning every 26 million years and bringing with it each time, by gravitational pull, a murderous rain of interplanetary debris that devastates ecosystems and wipes out many species. Sepkoski's brief presentation of the idea was provocative enough to draw attention not just in *Science* and *Science News* but also in the *Los Angeles Times*. Some people labeled that invisible companion "Nemesis," after the Greek goddess of vengeance, and others casually called it "the Death Star." Meanwhile, another of the presentations at Flagstaff was attracting no such fuss, but that's the one Jablonski remembers now.

It was a talk by Daniel Simberloff, an ecologist then at Florida State University, highly respected for his incisive mind but notoriously reluctant to draw sweeping conclusions from limited data. Simberloff's remarks carried the title "Are We on the Verge of a Mass Extinction in Tropical Rain Forests?" His answer, painstakingly reached, was yes.

"That's a really important paper, and a scary one," Jablonski says.

He vividly recalls the Flagstaff situation. "It wasn't a media event. It was scientists talking to scientists, being very up-front about what the uncertainties were and what the problems were." The problems were forest destruction, forest fragmentation, the loss of species that follows from those factors, and the cascading additional extinctions that come when ecosystems unravel. The uncertainties were considerable too, since there is no positive evidence left behind, no corpus delicti, when a species of rare bird or unknown beetle disappears as a consequence of the incineration of its habitat. Proving a negative fact is always difficult, and extinction is inherently a negative fact: *Such-and-such no longer exists.* Some biologists had begun warning of an extinction crisis that would be epochal in scale, their concern based on inference from the destruction of habitats that harbor vast numbers of highly localized species—in particular, tropical forests—and a few of those biologists had vivified their warnings with numerical estimates. Simberloff set himself to a skeptic's question: Is the situation really so dire? From his own cautious inferences and extrapolations, he reported that "even with an increase in the rate of destruction, there is not likely to be a mass extinction by the end of the century comparable to those of the geological past." He meant the twentieth century, of course. But was he saying that the alarms were illusory? Simberloff's reputation was such that no one could doubt he would make any unfashionable, spoilsport pronouncement to which the data, or lack of them, led him. Instead he added that in the next century, "if there are no major changes in the way forests are treated, things may get much worse." His calculation suggested that if tropical forests in the Americas were reduced to what's presently set aside as parks and reserves, 66 percent of all the native plant species would disappear by the end of the twenty-first century, and 69 percent of all Amazon birds. Yes, it would be a catastrophe on the same scale as every mass extinction except the Permian, Simberloff concluded.

"For me," David Jablonski says now, "that was a turning point."

But it's not the *starting* point I asked about. By the time of the Flagstaff meeting, I remind him, the idea of convening biologists together with paleontologists for a discussion of mass extinction was almost obvious, as reflected by the fact that two such events occurred that year. When was the idea less obvious? When was it just a fresh, counterintuitive notion? Jablonski obliges me by pushing his memory a little harder—back, as it turns out, to his own work during graduate school.

In the 1960s and early 1970s, concern about human-caused extinctions was neither widespread nor ecologically astute. Some writers warned about "vanishing wildlife" and "endangered species," but generally the warnings were framed around individual species with popular appeal, such as the whooping crane, the tiger, the blue whale, the peregrine falcon. Back in 1958, the pioneering British ecologist Charles Elton had published a farsighted book about biological dislocations, *The Ecology of Invasions by Animals and Plants*; Rachel Carson in 1962, with *Silent Spring*, had alerted people to the widespread, pernicious effects of pesticides such as DDT; and David Ehrenfeld's *Biological Conservation* appeared in 1970. But those three were untypical in their grasp of larger contexts. During the 1970s a new form of concern broke forth—call it wholesale concern—from the awareness that unnumbered millions of narrowly endemic (that is, unique and localized) species inhabit the tropical forests and that those forests were quickly being cut. The World Wildlife Fund and the Smithsonian Institution sponsored a symposium in 1974 on the subject of biological impoverishment; the chief scientist at WWF and the main organizer of that event was a young ecologist named Thomas E. Lovejoy, not long removed from his own doctoral work on Amazon birds. Another early voice belonged to Norman Myers, a Berkeley-trained biologist based in Nairobi. In 1976, Myers published a paper in *Science* recommending greater attention to the economic pressures that drive habitat destruction and the consequent loss of species;

in passing, he also compared current extinctions with the rate during what he loosely called "the 'great dying' of the dinosaurs." David Jablonski, then a graduate student struggling to do his dissertation and pay his bills, read Myers's paper and tucked a copy into his files. The comparison to the Cretaceous extinction, an event about which he was knowledgeable, didn't seem to him incongruous. Soon afterward, in early 1978, Jablonski was running out of cash and so "finagled the opportunity" to offer a seminar course, a special elective for undergraduates, through one of the Yale residential colleges. "I decided to teach it on extinction," he says.

Now suddenly energized by this recollection, Jablonski dodges among his paper-pile stalagmites to a cabinet and returns with a twenty-year-old file. He flips through it, mesmerized like an old athlete over a scrapbook from his improbable youth. The yellowing sheets tell us that his course ran in autumn 1978 as college seminar 130a, "Crises in the Evolution of Life." Eleven weeks of class were devoted to paleontological fundamentals such as deep time, uniformitarian change, the tempo and mode of evolution, Darwin and Lamarck, Cuvier and Lyell, and then to signal episodes such as the Permian extinction, the Devonian extinction, the K-T event. Week twelve would connect paleontology with neontology. On that Tuesday evening, according to a typed outline saved in the old file, students would consider the past and future impact of *Homo sapiens*, concerning notably: "The diminution of global biotic diversity, and how (or if) it should be maintained. Climatic effects of human activities. Are we on the brink of a mass extinction? The past as the key to the present." It was the first class that David Jablonski ever taught.

Norman Myers's early role in this matter was important from several angles. "He was the guy who really started the quantification of extinction," Jablonski recalls. "Norman was a pretty lonely guy for a long time on that." In 1979, Myers published *The Sinking Ark*, which explained the extinction problem for a popular audience, and in 1980 he produced a report for the National Academy of Sciences, drily titled *Conversion of Tropical Moist Forests* but full of eloquent data tracing the worldwide destruction of rainforest ecosystems. In the former book, he offered some rough numbers and offhand projections. Between the years 1600 and 1900, by his count, humanity had caused the extinction of about 75 known species, almost all of them mammals and birds. Between 1900 and 1979, humans had extinguished another 75 known species. Repeating what he had said in *Science*, Myers noted that this provisional tally—totaling 150 known species, all lost in less than four centuries—was well above the rate of known losses during the Cretaceous extinction. But more worrisome was the inferable rate of unrecorded extinctions, recent and now impending, among tropical plants and animals still unidentified by science. He guessed that 25,000 plant species presently stood jeopardized, and maybe hundreds of thousands of insects. "By the late 1980s we could be facing a situation where one species becomes extinct each hour. By the time human communities establish ecologically sound life-styles, the fallout of species could total several million." Rereading those sentences now, I'm struck by the reckless optimism of his assumption that human communities eventually *will* establish "ecologically sound life-styles." But back in 1981, when I first encountered Myers's book, his predictions seemed shocking and gloomy.

A year after *The Sinking Ark* appeared, Tom Lovejoy of WWF offered his own cautionary guesstimate in a section of the *Global 2000* report to outgoing president Jimmy Carter. Based on current projections of forest loss and a plausible relationship between forest area and endemism, Lovejoy suggested that 15 to 20 percent of all species—amounting to millions—might be lost by the end of the twentieth century. In the course of his discussion, Lovejoy also coined a new phrase, "biological diversity," which seems obvious in retrospect but hadn't yet been in use for denoting the aggregate of what was at stake. The portmanteau version, "biodiversity," would be buckled together a few years later. Among field biologists, a sense of focused concern was taking hold.

These early tries at quantification proved consequential for two reasons. First, Myers and Lovejoy helped galvanize public concern over the seemingly abstract matter of how many species may be lost as humanity claims an ever larger share of Earth's landscape and resources. Second, the Myers and Lovejoy warnings became targets for a handful of critics, who used the inexactitude of those numbers to cast doubt on the reality of the whole problem. Most conspicuous among the naysayers was Julian Simon, an economist at the University of Maryland, who argued bullishly that human population growth and human resourcefulness would solve all problems worth solving, of which a decline in diversity of tropical insects wasn't one.

In a 1986 issue of *New Scientist*, Simon rebutted Norman Myers, based on his own construal of select data, to the effect that there was "no obvious recent downward trend in world forests—no obvious 'losses' at all, and certainly no 'near catastrophic' loss." He later coauthored an op-ed piece in the *New York Times* under the headline "Facts, Not Species, Are Periled." Again he went after Myers, asserting a "complete absence of evidence for the claim that the extinction of species is going up rapidly—or even going up at all." Simon's worst disservice to logic in that statement and others was the denial that *inferential* evidence of wholesale extinction counts for anything. Of inferential evidence there was an abundance—for example, from the Centinela Ridge in a cloud-forest zone of western Ecuador, where in 1978 the botanist Alwyn Gentry and a colleague found thirty-eight species of narrowly endemic plants, including several with mysterious black leaves. Before Gentry could get back, Centinela Ridge had been completely deforested, the native plants replaced by cacao and other crops. As for inferential evidence generally, we might do well to remember what it contributes to our conviction that approximately 105,000 Japanese civilians died in the atomic bombing of Hiroshima. The city's population fell abruptly on August 6, 1945, but there was no one-by-one identification of 105,000 bodies.

Nowadays a few younger writers have taken Simon's line, pooh-poohing the concern over extinction. As for Simon himself, who died in 1998, perhaps the truest sentence he left behind was "We must also try to get more reliable information about the number of species that might be lost with various changes in the forests." No one could argue.

But it isn't easy to get such information. Field biologists tend to avoid investing their precious research time in doomed tracts of forest. Beyond that, our culture offers little institutional support for the study of narrowly endemic species in order to register their existence *before* their habitats are destroyed. Despite these obstacles, recent efforts to quantify rates of extinction have supplanted the old warnings. These new estimates use satellite imaging and improved on-the-ground data about deforestation, records of the many human-caused extinctions on islands, and a branch of ecological theory called island biogeography, which connects documented island cases with the mainland problem of forest fragmentation. These efforts differ in particulars, reflecting how much uncertainty is still involved, but their varied tones form a chorus of consensus. I'll mention three of the most credible.

W. V. Reid, of the World Resources Institute, in 1992 gathered numbers on the average annual deforestation in each of sixty-three tropical countries during the 1980s, and from them he charted three different scenarios (low, middle, high) of presumable forest loss by the year 2040. He chose a standard mathematical model of the relationship between decreasing habitat area and decreasing species diversity, made conservative assumptions about the crucial constant, and ran his various deforestation estimates through the model. Reid's calculations suggest that by the year 2040, between 17 and 35 percent of tropical forest species will be extinct or doomed to extinction. At either the high or the low end of this range, it would amount to a bad loss, though not as bad as the K-T event. Then again, 2040 won't mark the end of human pressures on biological diversity or landscape.

Robert M. May, an ecologist at Oxford, coauthored a similar effort in 1995. May and his colleagues noted the five causal factors that account for most extinctions: habitat destruction, habitat fragmentation, overkill, invasive species, and secondary effects cascading through an ecosystem from other extinctions. Each of those five is more intricate than it sounds. For instance, habitat fragmentation dooms species by consigning them to small parcels of habitat left insularized in an ocean of human impact and by then subjecting them to the same jeopardies (small population size, acted upon by environmental fluctuation, catastrophe, inbreeding, bad luck, and cascading effects) that make island species especially vulnerable to extinction. May's team concluded that most extant bird and mammal species can expect average life spans of between two hundred and four hundred years. That's equivalent to saying that about a third of one percent will go extinct each year until some unimaginable end point is reached. "Much of the diversity we inherited," May and his coauthors wrote, "will be gone before humanity sorts itself out."

The most recent estimate comes from Stuart L. Pimm and Thomas M. Brooks, ecologists at the University of Tennessee. Using a combination of published data on bird species lost from forest fragments and field data they gathered themselves, Pimm and Brooks concluded that 50 percent of the world's forest bird species will be doomed to extinction by deforestation occurring over the next half-century. And birds won't be the sole victims. "How many species will be lost if current trends continue?" the two scientists asked. "Somewhere between one third and two thirds of all species—easily making this event as large as the previous five mass extinctions the planet has experienced."

Jablonski, who started down this line of thought in 1978, offers me a reminder about the conceptual machinery behind such estimates. "All mathematical models," he says cheerily, "are wrong. They are approximations. And the question is: Are they usefully wrong, or are they meaninglessly wrong?" Models projecting present and future species loss are useful, he suggests, if they help people realize that *Homo sapiens* is perturbing Earth's biosphere to a degree it hasn't often been perturbed before. In other words, that this is a drastic experiment in biological drawdown we're engaged in, not a continuation of routine.

Behind the projections of species loss lurk a number of critical but hard-to-plot variables, among which two are especially weighty: continuing landscape conversion and the growth of human population.

Landscape conversion can mean many things: draining wetlands to build roads and airports, turning tallgrass prairies under the plow, fencing savanna and overgrazing it with domestic stock, cutting second-growth forest in Vermont and consigning the land to ski resorts or vacation suburbs, slash-and-burn clearing of Madagascar's rainforest to grow rice on wet hillsides, industrial logging in Borneo to meet Japanese plywood demands. The ecologist John Terborgh and a colleague, Carel P. van Schaik, have described a four-stage process of landscape conversion that they call the land-use cascade. The successive stages are: 1) *wildlands*, encompassing native floral and faunal communities altered little or not at all by human impact; 2) *extensively used areas*, such as natural grasslands lightly grazed, savanna kept open for prey animals by infrequent human-set fires, or forests sparsely worked by slash-and-burn farmers at low density; 3) *intensively used areas*, meaning crop fields, plantations, village commons, travel corridors, urban and industrial zones; and finally 4) *degraded land*, formerly useful but now abused beyond value to anybody. Madagascar, again, would be a good place to see all four stages, especially the terminal one. Along a thin road that leads inland from a town called Mahajanga, on the west coast, you can gaze out over a vista of degraded land—chalky red hills and gullies, bare of forest, burned too often by graziers wanting a short-term burst of pasturage, sparsely covered in dry grass and scrubby fan palms, eroded starkly, draining red mud into the Betsiboka River, supporting almost no human presence. Another showcase of degraded land—attributable to fuel-wood gathering, overgrazing, population density, and decades of apartheid—is the Ciskei homeland in South Africa. Or you might look at overirrigated crop fields left ruinously salinized in the Central Valley of California.

Among all forms of landscape conversion, pushing tropical forest from the wildlands category to the intensively used category has the greatest impact on biological diversity. You can see it in western India, where a spectacular deciduous ecosystem known as the Gir Forest (home to the last surviving population of the Asiatic lion, *Panthera leo persica*) is yielding along its ragged edges to new mango orchards, peanut fields, and lime quarries for cement. You can see it in the central Amazon, where big tracts of rainforest have been felled and burned, in a largely futile attempt (encouraged by misguided government incentives, now revoked) to pasture cattle on sun-hardened clay. According to the United Nations Food and Agriculture Organization, the rate of deforestation in tropical countries has increased (contrary to Julian Simon's claim) since the 1970s, when Norman Myers made his estimates. During the 1980s, as the FAO reported in 1993, that rate reached 15.4 hectares (a hectare being the metric equivalent of 2.5 acres) annually. South America was losing 6.2 million hectares of forest a year. Southeast Asia was losing less in sheer area but more proportionally: 1.6 percent of its forests yearly. In terms of cumulative loss, as reported by other observers, the Atlantic coast forest of Brazil is at least 95 percent gone. The Philippines, once nearly covered with rainforest, has lost 92 percent. Costa Rica has continued to lose forest, despite that country's famous concern for its biological resources. The richest old-growth lowland forests in West Africa, India, the Greater Antilles, Madagascar, and elsewhere have been reduced to less than a tenth of their original areas. By the middle of the twenty-first century, if those trends continue, tropical forest will exist virtually nowhere outside of protected areas—that is, national parks, wildlife refuges, and other official reserves.

How many protected areas will there be? The present worldwide total is about 9,800, encompassing 6.3 percent of the planet's land area. Will those parks and reserves retain their full biological diversity? No. Species with large territorial needs will be unable to maintain viable population levels within small reserves, and as those species die away, their absence will affect others. The disappearance of big predators, for instance, can release limits on medium-sized predators and scavengers, whose overabundance can drive still other species (such as ground-nesting birds) to extinction. This has already happened in some habitat fragments, such as Panama's Barro Colorado Island, and been well documented in the literature of island biogeography. The lesson of fragmented habitats is Yeatsian: Things fall apart.

Human population growth will make a bad situation worse by putting ever more pressure on all available land.

Population growth rates have declined in many countries within the past several decades, it's true. But world population is still increasing, and even if average fertility suddenly, magically dropped to 2.0 children per female, population would continue to increase (on the momentum of birthrate exceeding death rate among a generally younger and healthier populace) for some time. The annual increase is now 80 million people, with most of that increment coming in less developed countries. The latest long-range projections from the Population Division of the United Nations, released in early 1998, are slightly down from previous long-term projections in 1992 but still point toward a problematic future. According to the UN's middle estimate (and most probable? that's hard to know) among seven fertility scenarios, human population will rise from the present 5.9 billion to 9.4 billion by the year 2050, then to 10.8 billion by 2150, before leveling off there at the end of the twenty-second century. If it happens that way, about 9.7 billion people will inhabit the countries included within Africa, Latin America, the Caribbean, and Asia. The total population of those countries—most of which are in the low latitudes, many of which are less developed, and which together encompass a large portion of Earth's remaining tropical forest—will be more than twice what it is today. Those 9.7 billion people, crowded together in hot places, forming the ocean within which tropical nature reserves are insularized, will constitute 90 percent of humanity. Anyone interested in the future of biological diversity needs to think about the pressures these people will face, and the pressures they will exert in return.

We also need to remember that the impact of *Homo sapiens* on the biosphere can't be measured simply in population figures. As the population expert Paul Harrison pointed out in his book *The Third Revolution*, that impact is a product of three variables: population size, consumption level, and technology. Although population growth is highest in less-developed countries, consumption levels are generally far higher in the developed world (for instance, the average American consumes about ten times as much energy as the average Chilean, and about a hundred times as much as the average Angolan) and also higher among the affluent minority in any country than among the rural poor. High consumption exacerbates the impact of a given population, whereas technological developments may either exacerbate it further (think of the automobile, the air conditioner, the chainsaw) or mitigate it (as when a technological innovation improves efficiency for an established function). All three variables play a role in every case, but a directional change in one form of human impact—on air pollution from fossil-fuel burning, say, or fish harvest from the seas—can be mainly attributable to a change in one variable, with only minor influence from the other two. Sulfur dioxide emissions from developed countries fell dramatically during the 1970s and '80s, owing to technological improvements in papermaking and other industrial processes; those emissions would have fallen still further if not for increased population (accounting for 25 percent of the upward vector) and increased consumption (accounting for 75 percent). Deforestation, in contrast, is a directional change that *has* been mostly attributable to population growth.

According to Harrison's calculations, population growth accounted for 79 percent of the deforestation in less developed countries between 1973 and 1988. Some experts would argue with those calculations, no doubt, and insist on redirecting our concern toward the role that distant consumers, wood-products buyers among slow-growing but affluent populations of the developed nations, play in driving the destruction of Borneo's dipterocarp forests or the hardwoods of West Africa. Still, Harrison's figures point toward an undeniable reality: More total people will need more total land. By his estimate, the minimum land necessary for food growing and other human needs (such as water supply and waste dumping) amounts to one fifth of a hectare per person. Given the UN's projected increase of 4.9 billion souls before the human population finally levels off, that comes to another billion hectares of human-claimed landscape, a billion hectares less forest—even without allowing for any further deforestation by the current human population or for any further loss of agricultural land to degradation. A billion hectares—in other words, 10 million square kilometers—is, by a conservative estimate, well more than half the remaining forest area in Africa, Latin America, and Asia.

This raises the vision of a very exigent human population pressing snugly around whatever patches of natural landscape remain.

Add to that vision the extra, incendiary aggravation of poverty. According to a recent World Bank estimate, about 30 percent of the total population of less developed countries lives in poverty. Alan Durning, in his 1992 book *How Much Is Enough? The Consumer Society and the Fate of the Earth*, puts it in a broader perspective when he says that the world's human population is divided among three "ecological classes": the consumers, the middle-income, and the poor. His consumer class includes those 1.1 billion fortunate people whose annual income per family member is more than $7,500. At the other extreme, the world's poor also number about 1.1 billion people—all from households with less than $700 annually per family member. "They are mostly rural Africans, Indians, and other South Asians," Durning writes. "They eat almost exclusively grains, root crops, beans, and other legumes, and they drink mostly unclean water. They live in huts and shanties, they travel by foot, and most of their possessions are constructed of stone, wood, and other substances available from the local environment." He calls them the "absolute poor." It's only reasonable to assume that another billion people will be added to that class, mostly in what are now the less-developed countries, before population growth stabilizes. How will those additional billion, deprived of education and other advantages, interact with the tropical landscape? Not likely by entering information-intensive jobs in the service sector of the new global economy. Julian Simon argued that human ingenuity—and, by extension, human population itself—is "the ultimate resource" for solving Earth's problems, transcending Earth's limits, and turning scarcity into abundance. But if all the bright ideas generated by a human population of 5.9 billion haven't yet relieved the desperate needfulness of the 1.1 billion absolute poor, why should we expect that human ingenuity will do any better for roughly 2 billion poor in the future?

Other writers besides Durning have warned about this deepening class rift. Tom Athanasiou, in *Divided Planet: The Ecology of Rich and Poor*, sees population growth only exacerbating the division, and notes that governments often promote destructive schemes of transmigration and rainforest colonization as safety valves for the pressures of land hunger and discontent. A young Canadian policy analyst named Thomas Homer-Dixon, the author of several calm-voiced but frightening articles on the linkage between what he terms "environmental scarcity" and global sociopolitical instability, reports that the amount of cropland available per person is falling in the less developed countries because of population growth and because millions of hectares "are being lost each year to a combination of problems, including encroachment by cities, erosion, depletion of nutrients, acidification, compacting and salinization and waterlogging from overirrigation." In the cropland pinch and other forms of environmental scarcity, Homer-Dixon foresees potential for "a widening gap" of two sorts—between demands on the state and its ability to deliver, and more basically between rich and poor. In conversation with the journalist Robert D. Kaplan, as quoted in Kaplan's book *The Ends of the Earth*, Homer-Dixon said it more vividly: "Think of a stretch limo in the potholed streets of New York City, where homeless beggars live. Inside the limo are the air-conditioned post-industrial regions of North America, Europe, the emerging Pacific Rim, and a few other isolated places, with their trade summitry and computer information highways. Outside is the rest of mankind, going in a completely different direction." That direction, necessarily, will be toward ever more desperate exploitation of landscape. Kaplan himself commented: "We are entering a bifurcated world."

H. G. Wells foretold that bifurcation a century ago in his novel *The Time Machine*. Wells's time traveler, bouncing forward from Victorian London to the year A.D. 802,701, found a divided planet too, upon which the human race had split into two very different forms: the groveling, dangerous Morlocks, who lived underground, and the epicene Eloi, who enjoyed lives of languid comfort on the surface. The only quaint thing about Wells's futurology, from where we sit now, is that he imagined it would be necessary to travel so far.

As for Homer-Dixon's vehicle: When you think of that stretch limo on those potholed urban streets, don't assume there will be room inside for tropical forests. Even Noah's ark managed to rescue only paired animals, not large parcels of habitat. The jeopardy of the ecological fragments that we presently cherish as parks, refuges, and reserves is already severe, due to both internal and external forces: internal, because insularity leads to ecological unraveling; and external, because those areas are still under siege by needy and covetous people. Projected forward into a future of 10.8 billion humans, of which perhaps 2 billion are starving at the periphery of those areas, while another 2 billion are living in a fool's paradise maintained by unremitting exploitation of whatever resources remain, that jeopardy increases to the point of impossibility. In addition, any form of climate change in the midterm future, whether caused by greenhouse gases or by the natural flip-flop of climatic forces, is liable to change habitat conditions within a given protected area beyond the tolerance range for many species. If such creatures can't migrate beyond the park or reserve boundaries in order to chase their habitat needs, they may be "protected" from guns and chainsaws within their little island, but they'll still die.

We shouldn't take comfort in assuming that at least Yellowstone National Park will still harbor grizzly bears in the year 2150, that at least Royal Chitwan in Nepal will still harbor tigers, that at least Serengeti in Tanzania and Gir in India will still harbor lions. Those predator populations, and other species down the cascade, are likely to disappear. "Wildness" will be a word applicable only to urban turmoil. Lions, tigers, and bears will exist in zoos, period. Nature won't come to an end, but it will look very different.

The most obvious differences will be those I've already mentioned: tropical forests and other terrestrial ecosystems will be dramatically reduced in area, and the fragmented remnants will stand tiny and isolated. Because of those two factors, plus the cascading secondary effects, plus an additional dire factor I'll mention in a moment, much of Earth's biological diversity will be gone. How much? That's impossible to predict confidently, but the careful guesses of Robert May, Stuart Pimm, and other biologists suggest losses reaching half to two thirds of all species. In the oceans, deepwater fish and shellfish populations will be drastically depleted by over-harvesting, if not to the point of extinction, then at least enough to cause more cascading consequences. Coral reefs and other shallow-water ecosystems will be badly stressed, if not devastated, by erosion and chemical runoff from the land. The additional dire factor is invasive species, the fifth of the five factors contributing to our current experiment in mass extinction.

That factor, even more than habitat destruction and fragmentation, is a symptom of modernity. Maybe you haven't heard much about invasive species, but in coming years you will. Daniel Simberloff, the same ecologist who gave that sobering paper that Jablonski remembers from 1983, takes it so seriously that he recently committed himself to founding an institute on invasive biology at the University of Tennessee, and Interior Secretary Bruce Babbitt sounded the alarm in April 1998 in a speech to a weed-management symposium in Denver. The spectacle of a cabinet secretary denouncing an alien plant called purple loosestrife struck some observers as droll, but it wasn't as silly as it seemed. Forty years ago, Charles Elton warned in *The Ecology of Invasions by Animals and Plants* that "we are living in a period of the world's history when the mingling of thousands of kinds of organisms from different parts of the world is setting up terrific dislocations in nature." Elton's word "dislocations" was nicely chosen to ring with a double meaning: Species are being moved from one location to another, and as a result ecosystems are being thrown into disorder.

The problem dates back to when people began using ingenious new modes of conveyance (the horse, the camel, the canoe) to travel quickly across mountains, deserts, and oceans, taking with them rats, lice, disease microbes, burrs, dogs, pigs, goats, cats, cows, and other forms of parasitic, commensal, or domesticated creature. One immediate result of those travels was a wave of island-bird extinctions, claiming more than a thousand species, which followed oceangoing canoes across the Pacific and elsewhere. Having evolved in insular ecosystems free of predators, many of those species were flightless, unequipped to defend themselves or their eggs against ravenous mammals. *Raphus cucullatus*, a giant cousin of the pigeon lineage, endemic to Mauritius in the Indian Ocean and better known as the dodo, was only the most easily caricatured representative of this much larger pattern.

Dutch sailors killed and ate dodos during the seventeenth century, but probably what guaranteed the extinction of *Raphus cucullatus* is that the European ships put ashore rats, pigs, and *Macaca fascicularis*, an opportunistic species of Asian monkey. Although commonly known as the crab-eating macaque, *M. fascicularis* will eat almost anything. The monkeys are still pestilential on Mauritius, hungry and daring and always ready to grab what they can, including raw eggs. But the dodo hasn't been seen since 1662.

The European age of discovery and conquest was also the great age of biogeography—that is, the study of what creatures live where, a branch of biology practiced by attentive travelers such as Carl Linnaeus, Alexander von Humboldt, Charles Darwin, and Alfred Russel Wallace. Darwin and Wallace even made biogeography the basis of their discovery that species, rather than being created and plopped onto Earth by divine magic, evolve in particular locales by the process of natural selection. Ironically, the same trend of far-flung human travel that gave biogeographers their data also began to muddle and nullify those data, by transplanting the most ready and roguish species to new places and thereby delivering misery unto death for many other species. Rats and cats went everywhere, causing havoc in what for millions of years had been sheltered, less competitive ecosystems. The Asiatic chestnut blight and the European starling came to America; the American muskrat and the Chinese mitten crab got to Europe. Sometimes these human-mediated transfers were unintentional, sometimes merely shortsighted. Nostalgic sportsmen in New Zealand imported British red deer; European brown trout and coastal rainbows were planted in disregard of the native cutthroat trout of Rocky Mountain rivers. Prickly-pear cactus, rabbits, and cane toads were inadvisedly welcomed to Australia. Goats went wild in the Galápagos. The bacteria that cause bubonic plague journeyed from China to Europe by way of fleas, rats, Mongolian horsemen, and sailing ships, and eventually traveled also to California. The Atlantic sea lamprey found its own way up into Lake Erie, but only after the Welland Canal gave it a bypass around Niagara Falls. Unintentional or otherwise, all these transfers had unforeseen consequences, which in many cases included the extinction of less competitive, less opportunistic native species. The rosy wolfsnail, a small creature introduced onto Oahu for the purpose of controlling a larger and more obviously noxious species of snail, which was itself invasive, proved to be medicine worse than the disease; it became a fearsome predator upon native snails, of which twenty species are now gone. The Nile perch, a big predatory fish introduced into Lake Victoria in 1962 because it promised good eating, seems to have exterminated at least eighty species of smaller cichlid fishes that were native to the lake's Mwanza Gulf.

The problem is vastly amplified by modern shipping and air transport, which are quick and capacious enough to allow many more kinds of organism to get themselves transplanted into zones of habitat they never could have reached on their own. The brown tree snake, having hitchhiked aboard military planes from the New Guinea region near the end of World War II, has eaten most of the native forest birds of Guam. The same virus that causes monkeypox among Congolese villagers traveled to Wisconsin by way of certain African rodents, which were imported for the exotic wildlife trade; the virus then crossed into captive American prairie dogs and from them into people who thought prairie dogs would make nifty pets. SARS rode from Hong Kong to Toronto as the respiratory distress of one airline passenger. Ebola will next appear who knows where. Apart from the frightening epidemiological possibilities, agricultural damages are the most conspicuous form of impact. One study, by the congressional Office of Technology Assessment, reports that in the United States, 4,500 nonnative species have established free-living populations, of which about 15 percent cause severe harm; looking at just seventy-nine of those species, the OTA documented $97 billion in damages. The lost value in Hawaiian snail species or cichlid diversity is harder to measure. But another report, from the UN Environmental Program, declares that almost 20 percent of the world's endangered vertebrates suffer from pressures (competition, predation, habitat transformation) created by exotic interlopers. Michael Soulé, a biologist much respected for his work on landscape conversion and extinction, has said that invasive species may soon surpass habitat loss and fragmentation as the major cause of "ecological disintegration." Having exterminated Guam's avifauna, the brown tree snake has lately been spotted in Hawaii.

Is there a larger pattern to these invasions? What do fire ants, zebra mussels, Asian gypsy moths, tamarisk trees, maleleuca trees, kudzu, Mediterranean fruit flies, boll weevils, and water hyacinths have in common with crab-eating macaques or Nile perch? Answer: They are *weedy* species, in the sense that animals as well as plants can be weedy. What that implies is a constellation of characteristics: They reproduce quickly, disperse widely when given a chance, tolerate a fairly broad range of habitat conditions, take hold in strange places, succeed especially well in disturbed ecosystems, and resist eradication once they're established. They are scrappers, generalists, opportunists. They tend to thrive in human-dominated terrain because in crucial ways they resemble *Homo sapiens*: aggressive, versatile, prolific, and ready to travel. The city pigeon, a cosmopolitan creature derived from wild ancestry as a Eurasian rock dove (*Columba livia*) by way of centuries of pigeon fanciers, whose coop-bred birds occasionally went AWOL, is a weed. So are those species that, benefiting from human impacts upon landscape, have increased grossly in abundance or expanded their geographical scope without having to cross an ocean by plane or by boat—for instance, the coyote in New York, the raccoon in Montana, the whitetail deer in northern Wisconsin or western Connecticut. The brown-headed cowbird, also weedy, has enlarged its range from the eastern United States into the agricultural Midwest at the expense of migratory songbirds. In gardening usage, the word "weed" may be utterly subjective, indicating any plant you don't happen to like, but in ecological usage it has these firmer meanings. Biologists frequently talk of weedy species, referring to animals as well as plants.

Paleontologists too embrace the idea and even the term. Jablonski himself, in a 1991 paper published in *Science*, extrapolated from past mass extinctions to our current one and suggested that human activities are likely to take their heaviest toll on narrowly endemic species, while causing fewer extinctions among those species that are broadly adapted and broadly distributed. "In the face of ongoing habitat alteration and fragmentation," he wrote, "this implies a biota increasingly enriched in widespread, weedy species—rats, ragweed, and cockroaches—relative to the larger number of species that are more vulnerable and potentially more useful to humans as food, medicine, and genetic resources." Now, as we sit in his office, he repeats: "It's just a question of how much the world becomes enriched in these weedy species." Both in print and in talk he uses "enriched" somewhat caustically, knowing that the actual direction of the trend is toward impoverishment of variety.

Regarding impoverishment, let's note another dark, interesting irony: that the two converse trends I've described—partitioning the world's landscape by habitat fragmentation and unifying the world's landscape by global transport of weedy species—produce not converse results but one redoubled result, the further loss of biological diversity. Immersing myself in the literature of extinctions and making dilettantish excursions across India, Madagascar, New Guinea, Indonesia, Brazil, Guam, Australia, New Zealand, Wyoming, the hills of Burbank, and other semiwild places over the past decade, I've seen those redoubling trends everywhere, portending a near-term future in which Earth's landscape is threadbare, leached of diversity, heavy with humans, and "enriched" in weedy species. That's an ugly vision, but I find it vivid. Wildlife will consist of the pigeons and the coyotes and the whitetails, the black rats (*Rattus rattus*) and the brown rats (*Rattus norvegicus*) and a few other species of worldly rodent, the crab-eating macaques and the cockroaches (though, as with the rats, not *every* species—some are narrowly endemic, like the giant Madagascar hissing cockroach) and the mongooses, the house sparrows and the house geckos and the houseflies and the barn cats and the skinny brown feral dogs and a short list of additional species that play by our rules. Forests will be tiny insular patches existing on bare sufferance, much of their biological diversity (the big predators, the migratory birds, the shy creatures that can't tolerate edges, and many other species linked inextricably with those) long since decayed away. They will essentially be tall woody gardens, not forests in the richer sense. Elsewhere the landscape will have its strips and swatches of green, but except on much-poisoned lawns and golf courses, the foliage will be infested with cheatgrass and European buckthorn and spotted knapweed and Russian thistle and leafy spurge and salt-meadow cordgrass and Bruce Babbitt's purple loosestrife. Having recently passed the great age of biogeography, we will have entered the age *after* biogeography, in that virtually everything will live virtually everywhere, though the list of species that constitute "everything" will be small. I see this world implicitly foretold in the UN population projections, the FAO reports on deforestation, the northward advance into Texas of Africanized honeybees, the rhesus monkeys that haunt the parapets of public buildings in New Delhi, and every fat gray squirrel on a bird feeder in England. Earth will be a different sort of place—soon, in just five or six human generations. My label for that place, that time, that apparently unavoidable prospect, is the Planet of Weeds. Its main consoling felicity, as far as I can imagine, is that there will be no shortage of crows.

Now we come to the question of human survival, a matter of some interest to many. We come to a certain fretful leap of logic that otherwise thoughtful observers seem willing, even eager, to make: that the ultimate consequence will be the extinction of us. By seizing such a huge share of Earth's landscape, by imposing so wantonly on its providence and presuming so recklessly on its forgiveness, by killing off so many species, they say, we will doom our own species to extinction. This is a commonplace argument among the environmentally exercised. In earlier years, from a somewhat less informed perspective, I made the same argument myself. Since then, my thinking has changed. My objection to the idea now is that it seems ecologically improbable and too optimistic. But it bears examining, because it's frequently offered as the ultimate argument against proceeding as we are.

Jablonski also has his doubts. Do you see *Homo sapiens* as a likely survivor, I ask him, or as a casualty? "Oh, we've got to be one of the most bomb-proof species on the planet," he says. "We're geographically widespread, we have a pretty remarkable reproductive rate, we're incredibly good at co-opting and monopolizing resources. I think it would take a really serious, concerted effort to wipe out the human species." The point he's making is one that has probably already dawned on you: *Homo sapiens* itself is the consummate weed. Why shouldn't we survive, then, on the Planet of Weeds? But there's a wide range of possible circumstances, Jablonski reminds me, between the extinction of our species and the continued growth of human population, consumption, and comfort. "I think we'll be one of the survivors," he says, "sort of picking through the rubble." Besides the loss of all the pharmaceutical and genetic resources that lay hidden within those extinguished species, and all the spiritual and aesthetic values they offered, he foresees unpredictable levels of loss in many physical and biochemical functions that ordinarily come as benefits from diverse, robust ecosystems—functions such as cleaning and recirculating air and water, mitigating droughts and floods, decomposing wastes, controlling erosion, creating new soil, pollinating crops, capturing and transporting nutrients, damping short-term temperature extremes and longer-term fluctuations of climate, restraining outbreaks of pestiferous species, and shielding Earth's surface from the full brunt of ultraviolet radiation. Strip away the ecosystems that perform those services, Jablonski says, and you can expect grievous detriment to the reality we inhabit. "A lot of things are going to happen that will make this a crummier place to live—a more stressful place to live, a more difficult place to live, a less resilient place to live—before the human species is at any risk at all." And maybe some of the new difficulties, he adds, will serve as incentive for major changes in the trajectory along which we pursue our aggregate self-interests. Maybe we'll pull back before our current episode matches the Triassic extinction or the K-T event. Maybe it will turn out to be no worse than the Eocene extinction, with a 35 percent loss of species.

"Are you hopeful?" I ask.

Given that hope is a duty from which paleontologists are exempt, I'm surprised when he answers, "Yes, I am."

I'm not. My own guess about the midterm future, excused by no exemption, is that our Planet of Weeds will indeed be a crummier place, a lonelier and uglier place, and a particularly wretched place for the 2 billion people composing Alan Durning's absolute poor. What will increase most dramatically as time proceeds, I suspect, won't be generalized misery or futuristic modes of consumption but the gulf between two global classes experiencing those extremes. Progressive failure of ecosystem functions? Yes, but human resourcefulness of the sort Julian Simon so admired will probably find stopgap technological remedies, to be available for a price. So the world's privileged class—that's your class and my class—will probably still manage to maintain themselves inside Homer-Dixon's stretch limo, drinking bottled water and breathing bottled air and eating reasonably healthy food that has become incredibly precious, while the potholes in the road outside grow ever deeper. Eventually the limo will look more like a lunar rover. Ragtag mobs of desperate souls will cling to its bumpers, like groupies on Elvis's final Cadillac. The absolute poor will suffer their lack of ecological privilege in the form of lowered life expectancy, bad health, absence of education, corrosive want, and anger. Maybe in time they'll find ways to gather themselves in localized revolt against the affluent class, and just set to eating them, as Wells's Morlocks ate the Eloi. Not likely, though, as long as affluence buys guns. In any case, well before that they will have burned the last stick of Bornean dipterocarp for firewood and roasted the last lemur, the last grizzly bear, the last elephant left unprotected outside a zoo.

Jablonski has a hundred things to do before leaving for Alaska, so after two hours I clear out. The heat on the sidewalk is fierce, though not nearly as fierce as this summer's heat in New Delhi or Dallas, where people are dying. Since my flight doesn't leave until early evening, I cab downtown and take refuge in a nouveau-Cajun restaurant near the river. Over a beer and jambalaya, I glance again at Jablonski's 1991 *Science* paper, titled "Extinctions: A Paleontological Perspective." I also play back the tape of our conversation, pressing my ear against the little recorder to hear it over the lunch-crowd noise.

Among the last questions I asked Jablonski was, What will happen *after* this mass extinction, assuming it proceeds to a worst-case scenario? If we destroy half or two thirds of all living species, how long will it take for evolution to fill the planet back up? "I don't know the answer to that," he said. "I'd rather not bottom out and see what happens next." In the journal paper he had hazarded that, based on fossil evidence in rock laid down atop the K-T event and others, the time required for full recovery might be 5 or 10 million years. From a paleontological perspective, that's fast. "Biotic recoveries after mass extinctions are geologically rapid but immensely prolonged on human time scales," he wrote. There was also the proviso, cited from another expert, that recovery might not begin until after the extinction-causing circumstances have disappeared. But in this case, of course, the circumstances won't likely disappear until *we* do.

Still, evolution never rests. It's happening right now, in weed patches all over the planet. I'm not presuming to alert you to the end of the world, the end of evolution, or the end of nature. What I've tried to describe here is not an absolute end but a very deep dip, a repeat point within a long, violent cycle. Species die, species arise. The relative pace of those two processes is what matters. Even rats and cockroaches are capable—given the requisite conditions; namely, habitat diversity and time—of speciation. And speciation brings new diversity. So we might reasonably imagine an Earth upon which, 10 million years after the extinction (or, alternatively, the drastic transformation) of *Homo sapiens*, wondrous forests are again filled with wondrous beasts. That's the good news.

The River Jumps
Over the Mountain

LIFE IS SHORT and the Grand Canyon is long, especially when you paddle your way down it in a kayak. From the put-in at Lees Ferry, not far below Glen Canyon Dam, the Colorado River winds 226 miles between walls of primordial rock to a take-out at Diamond Creek, on the Hualapai Indian Reservation, dropping through dozens of major rapids along the way. Beyond that is slightly more river, more canyon, but the urgency, the majestic ferocity, and the sense of otherworldly containment dissipate down there, as the canyon walls tilt back into rubble slopes of Sonoran desert vegetation and the water's awesome momentum dribbles out anticlimactically into flat, inert Lake Mead. The deep magic and adamantine power that make this particular canyon grander than all others on Earth lie in those upper 226 miles, between the launch point and Diamond Creek. My own little kayak, of stiff yellow plastic, is nine foot two.

Simple arithmetic tells me that I'll need to travel 130,271 boat-lengths from start to finish. It's a ratio conducive to humility.

On water like this, each boat-length of headway involves two paddle strokes and, through the more serious rapids, maybe a quick tactical brace to prevent being flipped upside down. The lovely thing about a whitewater kayak is that, far beyond any other sort of water craft, it offers maneuverability in exchange for vulnerability, a tradeoff that intensifies the boater's sense of intimate interaction with a river. Climbing into a snugly fitted kayak, wedging your butt between the hip pads, arching your knees up into the thigh braces, is more like buckling on skis than like boarding a vessel. This is a sporting tool, not just a mode of conveyance. Stability is achieved, not given. A whitewater kayak even differs drastically from a sea kayak—roughly to the degree, say, that riding a unicycle in the circus differs from pedaling a ten-speed across Nebraska. Offering so little inherent equilibrium, so many dimensions of surprise, it's therefore the perfect boat in which to explore the chaotic border zones between equilibria and disequilibria of a personal nature—which is what, for me anyway, this trip is about. I've recently been set wobbling by the end of what I'd thought was a very good, very permanent marriage. A descent of the Grand Canyon by kayak should be more robust and less piteous, I figure, than a midlife crisis.

We launch on a Tuesday in early September, with the days growing shorter but the sun still high enough at midafternoon to make the deepest canyon rocks radiate, into evening, like oven-fresh bread. There are sixteen of us, seven in kayaks, the rest as oarsmen or passengers on inflatable rafts, a motley assemblage of old friends and new acquaintances all centered on the trip's organizational leaders, Cyndi and Bob Crayton of Bozeman, Montana.

Bob Crayton ran the Grand Canyon twenty years ago, as a young oil-field roughneck with a full head of hair, in a clumsy old fiberglass kayak that he paddled without undue concern about what lurked around the next bend, either on the river or in life. It was a larkish, bachelor getaway with a gang of male pals, yet he was so lingeringly affected by the experience that when he met Cyndi she took it to be something worth sharing. After the birth of their second child, she applied for a Grand Canyon permit herself. The responsible officials at Grand Canyon National Park allow only eight private-party launches per week, and the waiting list is lengthy. Eleven years later, Cyndi's name has come up, and her family—now including a lanky, handsome sixteen-year-old son, Chase, and a vivaciously feisty twelve-year-old daughter, Kinsey—forms the nucleus of our expeditionary party. Because the Park Service paperwork designates the permitee as "Trip Leader," and because she has borne so much of the organizational burden, we have all stopped calling Cyndi by her name and switched to the honorific title TL. Where are we camping tonight? Ask TL. Which box, on which raft, has the Pringles? Ask TL. Hey, TL, thanks for the margaritas! Rising to this burdensome challenge, Cyndi will eventually take to wearing a rhinestone tiara (belonging to Kinsey, who packed it for Mom as a surprise) on select occasions, when asserting her authority.

By terms of the permit, we have eighteen days to cover our 226 miles. Life is short and eighteen days still shorter—even if you're living out of boats, sleeping on the ground among scorpions and rattlesnakes, defecating into metal boxes, and bathing in a cold, silty river or not at all—but for an exercise in detachment from doleful confusions and mortal regrets, which is what I want, it should be sufficient.

An hour after launching, four miles downstream, we pass under Navajo Bridge, far above, and get our last glimpse for weeks of a vehicle that travels on wheels.

The river is slatey green and cold, having just emerged through dam gates from the bottom of Lake Powell. We paddle the flatwater stretches and bob through several warmup riffles, ogling the stone, elated that we're finally under way. The current slides along at about four miles per hour, and if we rode it passively, we could make a good day's distance between late morning and midafternoon, with no shortage of scenic amusements. Several bends downstream, we fall silent at the sight of several bighorn ewes grazing placidly on a sand flat along river right. They ignore us. A great blue heron roosts, with the cold dignity of a pterodactyl, on a high cliff. A belted kingfisher flies along one bank, making those trapeze-artist kingfisher swoops. The rock layers continue rising, revealing themselves as distinct strata and groups of strata by their differing colors and textures, and I've done just enough homework to try to identify them.

Let's see, from the top: *That* must be the Kaibab formation, then the Toroweap, then the Coconino sandstone just below. I can recite the cardinal sequence thanks to a mnemonic offered by a scientist friend before I left Montana: *Kissing Takes Concentration, However, Sex Requires More Breath And Tongue.* It codes for Kaibab, Toroweap, Coconino, Hermit shale, Supai group, Redwall limestone, Muav limestone, Bright Angel shale, and finally Tapeats sandstone. After a few dozen miles, I notice what seems to be a distinct formation—crumbly, rounded off, as red as dried blood— just emerging at river level. Rick Alexander, one of my kayaker pals, with a half-dozen previous Grand Canyon trips on his résumé and an appreciation for the place that goes beyond whitewater hydrology, confirms that we're now seeing the Hermit shale. I'm mesmerized by geologic spectacle, it's better than watching a lava lamp—but then *Kayaking Takes Concentration* too, and we climb out of our boats to scout Badger rapid.

My map booklet rates Badger's difficulty as 7 (on a scale of 10) at this water level, but it looks to be nothing more than a stairway of large, breaking waves, with a tongue of smooth green water marking the obvious line of entry. (Converted to the more standard scale of whitewater rating, Class I through Class V, the major Grand Canyon rapids could all be described as "Class IV, but big.") Rick's considered wisdom, after a glance, is "Hey diddle diddle, straight down the middle." And that's where we go.

Just below Badger, TL has decreed, is our camp spot for the night. We haven't covered much mileage, but never mind, we've consummated our escape from the realm of the dry. That we're just eight miles from the put-in is less relevant, suddenly, than that we're 218 miles from the take-out.

The moon appears late, as a waning gibbous shape over the south rim. The canyon walls occlude most of the sky, like big black shoulders, but along the linear gap between them stretches the Milky Way. So there's sky enough, stars enough, world enough and time, to lull even a full, busy, vexed mind to sleep. My own mind is weary and, as I've been hoping, empty.

The river is a pathway through rock. The rock is a pathway through time. The span of time manifest in the exposed rock of the Grand Canyon is vast almost behind comprehension, reflecting more than a third of the total age of our planet. The Vishnu schist, a steely gray metamorphic formation lining the innermost canyon gorge with polished cliffs that rise sheer from the water, dates back 1.7 billion years. The sedimentary layers lying on top are much younger, including that vertical stack of Paleozoic strata memorialized by my little mnemonic, all of which were laid down between 570 and 245 million years ago. That point bears emphasizing: that the *youngest* stratum atop the Grand Canyon rim derives from the end of the Paleozoic era, more than a quarter billion years ago. The Mesozoic era, with its giant reptiles, scarcely exists in this petrological record—too evanescent, too young. The Cenozoic, covering the past 65 million years, shows only as latter-day scuffs and scratches, such as the river canyon itself, or the spills of extruded lava that temporarily clogged it as recently as a mere million years ago. Among the more striking facts about this geological wonder—though not the single most mystifying one, which I'll come to in its turn—is that though the rock layers are extremely old, the canyon itself is quite recent. The river's channel (or at least the western half) seems to have been carved to nearly its present depth within just a few million years, and beginning only 6 million years ago. The river cut through like a silver knife slicing cake, though the cake itself had taken eons to assemble and bake.

My own age is fifty-three. That's risibly old on the kayaker scale and immeasurably young on the geologic one. Time is relative, Einstein taught us, and such relativity is another factor in my secret agenda of recuperation. Hey, Dave, we've got a Grand permit, Bob Crayton told me more than a year ago, want to come? Not possible, I thought—too many deadlines, too many commitments, it takes too much time, my kayak skills are in disrepair, I stand at the threshold of geezerhood, quack quack quack. And then, in a moment of sublime, reckless clarity, I said: *Yes!* No matter how old you are, I had realized, if you set yourself down within the ancientness of the Grand Canyon, your elapsed years will seem like nothing. Your life itself may seem like nothing. Your woes and your moans, your disappointments and sorrows and grievances and guilts, may therefore seem inconsiderable also. Rinse yourself in the river, measure yourself against the rock; find yourself to be a tiny, wet creature, insignificant within the larger and longer scope. That's the notion that put me on the trip roster. My shoulders are still in fairly good shape (always an issue, since dislocation is a common kayak injury) and, as far as diagnostic medicine can determine, so is my heart. I'll never know how old is *too* old, or not, unless I find out.

Among the seven of us paddling little boats, four are essentially professional kayakers, having grounded their lives in the sport either as instructors (Al Borrego, a quietly affable fellow who shifts to ski-patrolling in winter) or as sales reps (John Kudrna, Rob Lesser, and my geology consultant, Rick Alexander) within the whitewater world. Alexander, aka "Rick the Stick" for his paddling prowess, is a big burly guy roughly the size of a doorway, who looks like he might enjoy punishing people on a rugby field; his pale blue eyes and glinty smirk conceal—then sometimes reveal—a fundamental sweetness of character and a keen knowledge of natural history, gained during an earlier career directing outdoor-education programs in the Southwest. Kudrna is a compact white-water athlete whose shoulders have been rebuilt more times than the engine in a '64 Volkswagen. Lesser, with more than thirty years' paddling experience, is a legendary maker of first descents on harrowing Alaskan river canyons such as the Stikine. In addition, there's Mark Gamba, the photographer on assignment as my partner, whose long legs barely fit into a kayak. Mark's role obliges him to ballast the back of his boat with a case full of heavy photo gear and to wear a camera-encasing waterproof apparatus the size of a small television (he calls it a "surf housing") around his neck like a millstone. They're all younger and better paddlers than I— all except Lesser, who is older (God bless him) and (damn him) much better. Once again, as on previous kayak assignments, not all of which went off without ugly moments of drama, I find myself running in fast company.

The raft oarsmen include Brian Zimmer, a wry schoolteacher who consents cheerily to carry the army-surplus rocket boxes that will serve as receptacles for what is delicately known as our "groover" (the portable toilet) and accordingly christens his eighteen-foot yellow boat *Winnie da Pooh*; Jason Dzikowski, known as "Diz," a steady and earnest young carpenter from North Carolina, whose white raft is a twin to Brian's and therefore becomes *Piglet*; Mike Jaenish, a criminal defense attorney from Salt Lake City, soon to be a grandfather, who clears his court schedule and goes AWOL to run rivers at any reasonable offer, bringing his own raft (parakeet blue, with a banana-yellow sun canopy), his kitchen setup, his flask of Knob Creek bourbon, his aluminum cot, and his guitar; and Steve Jones, another lawyer by education but a contractor and a river rat by choice, whose renunciation of legal practice has allowed him, in the past three decades, to make fourteen previous Grand Canyon trips. These generous raft jockeys carry the freight that allows us kayakers, as well as themselves, to river-camp in comfort bordering on decadence: tents, lawn chairs, tables, beer, coolers full of fresh fruit and vegetables, frozen meat, many loaves of bread, many pounds of cheese, beer, rice, pasta, coffee, tortillas, canned beans, beer, dutch ovens, cookies, eight kinds of salsa, marshmallows, dozens of eggs, I think I've said beer, dry clothes, boccie balls, hiking shoes, two-burner propane stoves, a pancake griddle, a fire pan, charcoal briquets, battery-powered lanterns, some Budweiser for when the beer is all gone, and (I swear to God) a croquet set. Steve Jones has even thought to bring four pink plastic flamingos, with stab-in metal legs, for decorating the river frontage at each evening's camp.

Compared to an eighteen-foot raft loaded with such paraphernalia, a kayak has only the most modest capacity for cargo. It can carry a few items, but they had better be small and precious. As we launch on the morning of day two, and for every day thereafter, my own boat contains the following: an extra paddle, in two conjoinable pieces; air bags, to save the boat if I abandon ship and resort to swimming for my life; a pair of river sandals, for hiking side canyons; a water bottle; a rescue rope coiled in a throw bag; a baseball cap; a little waterproof pouch (which rides in a handy bungee-cord shelf under the front deck, just above my knees) containing my river map, a pencil, and a Rite in the Rain notebook; and a roll-top waterproof bag, holding certain important sundries. The sundries are my wallet, my watch, one energy bar, a container of sunscreen, and two books. The books are *Illustrations of the Huttonian Theory of the Earth*, by a Scottish mathematician named John Playfair, and *Selected Poems,* by W. H. Auden. I've chosen those two as my intellectual and emotional sustenance for the trip. Like the energy bar, they're small packets but densely nutritious.

I'll have no time to read except during stolen moments in camp, evening and morning, so the books could just as well be in my dry bag of other stuff (sleeping bag, pad, headlamp, etc.) lashed aboard *Piglet*. But I prefer keeping them close to me, like survival gear.

Playfair's *Illustrations of the Huttonian Theory of the Earth* is not just a classic of science explication but also a famous act of personal loyalty. James Hutton, another Scotsman, sometimes considered the founder of modern geology, conceived a revolutionary and percipient vision of how Earth's surface has been shaped and reshaped by geological processes. But the grand opus in which Hutton presented his ideas (*Theory of the Earth, with Proofs and Illustrations*, two volumes, 1795) was so turgid, so repetitious, and so poorly received, that his good friend Playfair undertook, after Hutton's death, to revivify the theory by describing it in concise, readable form. The essence of Hutton's theory centers on three points, all of which seemed outlandishly heterodox in his time: 1) Earth's surface is constantly being eroded by water, ice, and wind, which grind old rock into chunks, pebbles, and fine sediments that are carried downstream by rivers for eventual deposition on the sea bottoms; 2) sea-bottom sediments, transmogrified slowly by pressure and heat, become stratified layers of new rock; 3) further heat from below (what is its source?—that remained puzzling long after Hutton's time) also causes the slow uplift of those strata, and of the magmas of molten rock beneath them, eventually forming jagged mountains, domed plateaus, granitic knobs, great rifts and warpings, exposures and juxtapositions of variously tilted strata—all of which are subject to further erosion. In short, mountains become silt which becomes sedimentary rock which becomes mountains, with erosion driving the process from above and subterranean heat driving it from below, in a repeating cycle that seems to go on indefinitely, showing "no vestige of a beginning,—no prospect of an end."

Hutton wasn't an impious man, but his theory provoked accusations of impiety. Among its corollaries and saucy implications were that 4) marine fossils at high elevations were not put there by Noah's flood; 5) the processes affecting topography nowadays—erosion, deposition, barely detectable uplift, and an occasional volcanic burp—are the same and the only processes that shaped the world from its beginning; and therefore 6) planet Earth is much, much older than the figure of six-thousand-some years that had been calculated by biblical literalists. "Time," Hutton wrote, "which measures every thing in our idea, and is often deficient to our schemes, is to nature endless and as nothing." Most of Hutton's prose wasn't so piquant, and his friend Playfair did a breezier job of arguing the Huttonian case, describing great cycles of "decay and renovation" to account for the world as we see it. Everything that rises will be torn down, Playfair explained; everything torn down will be remade into something else, equally stony, equally grand, and rise again. My copy of Playfair's book is a facsimile of the first edition, published in 1802.

The Auden volume, by contrast, is a work of consummate twentieth-century modernity. Published in 1979, it samples the best of a long, vibrant poetic lifetime. Although a few of Auden's later poems are even more opaque than an eighteenth-century disquisition on geology, I find deep pleasure and consolation in his grim, mordant, yet bravely humane work from the 1930s. Some of it is political, some intimately personal. Certain of these poems are written in a deceptively simple style that flows like light verse. One that I've read often, and will read again on the river, begins this way:

> As I walked out one evening,
>   Walking down Bristol Street,
> The crowds upon the pavement
>   Were fields of harvest wheat.
>
> And down by the brimming river
>   I heard a lover sing
> Under an arch of the railway:
>   "Love has no ending.
>
> "I'll love you, dear, I'll love you
>   Till China and Africa meet
> And the river jumps over the mountain
>   And the salmon sing in the street."

Both books fit easily into the back of my boat, their waterproof bag clipped in with a carabiner, beside the rescue rope.

For most of its length in the canyon—say, 90 percent of those 226 miles—the Colorado River is like a giant sleeping snake, its latent power barely intimated by a gentle reptilian snore. The current glides slowly along, with only an occasional swell or whirlpool on an eddy line reflecting the vast, merciless energy held contained. But the river's elevation drops almost 2,000 feet between Lees Ferry and Lake Mead, and about half that total occurs in short, abrupt plunges—that is, rapids. During the next couple days, we get our first real taste of the river's wild side.

Soap Creek rapid is another frothy chain of waves, with no holes lurking to swallow and hold a boat, so Mark takes it as an opportunity for action close-ups. He runs Soap Creek backward, bracing himself with one hand, deploying his surf-housing camera with the other, clicking off motor-drive shots of Rob amid the churning jiggle-jaggle of the waves. Halfway through, Mark flips upside down. I watch his boat bottom from not far away, awaiting a recovery. Underwater, he drops the camera and gets both hands on his paddle, then rolls briskly up, to discover that one of his surf-housing straps has failed and the apparatus—all $3,400 of it, with a Nikon F-100 inside—seems to be gone. Bad moment, bad setback, to lose such a crucial piece of equipment so early in the trip. Then he notices the thing trailing behind him, on one strap, like a drag-bag of beer left in the water for chilling. He reels it in.

House Rock is more serious, a right-bending rapid in which the heavy chop pushes into a rock wall on the left and, near the bottom, a pair of tall waves guard the exit line, one of them not just breaking but recirculating. For the kayakers this doesn't present much trouble. Each of us enters on the flat green tongue, angling right, and with a few strokes amid the heavy water we're able to stay off the wall, ferry rightward, and punch through the wave-hole along its right corner. We catch eddies at the bottom and hold position, ready if needed to help with a rescue. For the rafts, so heavily loaded, so lumbering, it's a different matter. In fact, this particular rapid proves a good reminder of a truth we already know well:

Some stretches of water that are easily run by a competent kayaker can be wickedly problematic for a raft oarsman, and vice versa. In their strengths and their foibles, a big raft and a little kayak are as different as a locomotive and a horse.

Brian's locomotive runs next. With the unsavory groover boxes strapped firmly to its frame, with Chase Crayton and his high-school buddy Cole Arpin whooping in the bow, *Pooh* edges barely away from the wall, drops straight into the wave-hole, and goes nearly vertical, eighteen feet of fat yellow sausage standing on end. Diz follows the same line in *Piglet*, with young Kinsey Crayton and Margie Penney (a nurse from Colorado, old friend of Cyndi's) dangling forward, clutching handholds, to get thrillingly drenched in the breakers. Later, Bob will confide to me that he found himself quite flustered as he sculled in that eddy, watching his twelve-year-old daughter ride through the rapid. It was a new sort of whitewater excitement, jangling and unexpectedly disagreeable, for the old man. This time on the big river, he realized, he had given hostages to fortune.

And then comes Mike, his yellow canopy lowered for stormy running, a straw cowboy hat on his head. His raft being the lightest and the shortest, it's the most mobile but also the least stable, and he has no passengers to help with high-siding or bailing. He begins with an angle to the right, but then somehow his boat gets swung leftward, way leftward, and slides toward the paired waves like a van skidding on ice. When he hits the waves broadside, there's an alley-oop motion and Mike is suddenly in the air—then in the water, gone. He bobs up beside the raft, minus his hat, and catches hold of an oar. By the time kayaks converge on him, he has already hoisted himself back in and brought the boat under control, a nice recovery by any measure.

Mike has made five earlier Grand Canyon trips. Experienced and provident, he appears next day with a different hat.

By the end of a week we've followed the river downward through more than a billion years of time, descending past all nine major formations of Paleozoic sedimentary rock and into the Vishnu schist, dark and Precambrian. The cliff sides are suddenly closer, steeper, more stern and chilling, like melodramatic pinnacles in a woodcut by Rockwell Kent. The gunmetal-gray schist is shot through with sinuous veins of pinkish, mica-flecked intrusion, known as the Zoroaster granite. In their physical presence as well as their mythic evocations, the Vishnu and the Zoroaster provide a somber, eerie sense of embrace. Rick says: "Welcome to the inner canyon gorge."

I'm still wondering how the river carved its way down here so quickly. The question is made even more baffling by the geologic conundrum I alluded to earlier, which involves a mysterious surmounting of certain obstacles. "The Colorado River has cut through several major upwarps, including the Kaibab Plateau, seemingly in defiance of the laws of gravity," according to an expert named Larry Stevens. "Controversy over how and when the Grand Canyon formed has raged for a century, but every new theory seems to be missing a critical piece of evidence." One enigma any such theory must explain is that the early Colorado River, flowing at what seems to have been a middling elevation across an area known as the Marble Platform, managed to carve its way over—and then down into—a big, elongated dome of elevated rock known as the Kaibab Uplift. From the surface of the Marble Platform to the crest of the Kaibab Uplift, as they stand today, there's a *rise* of several thousand feet. Did the water run uphill? Certainly not. Then how did the river get over that mountainous mound?

Nobody knows. But three different hypotheses have been offered during the past century and a half, each bidding to explain it without recourse to miracles.

John Wesley Powell, after his explorations of the canyon in 1869 and 1872, guessed that the river had etched its path first, along what was a natural declivity, and that the Kaibab Uplift had risen afterward, raising the land surface against the river's flow like a loaf of bread being pushed into a band saw. Later research has discredited that guess by establishing that the river channel is more recent than the vaulting.

A second hypothesis, which held sway in the 1960s, was that the river essentially backed its way through the high ground of the canyon's middle reaches, by what geologists call "headward erosion." When a lump of rock is dislodged from the brink of, say, Niagara Falls, dropping into the gorge below, the brink itself recedes upstream by an increment equal to the size of the lost lump. That's headward erosion. The Canadian half of Niagara Falls, known as Horseshoe, is eroding headward at the speedy rate of about five feet per year. Moving just a fraction that fast, the Colorado might have eaten backward through the Kaibab Uplift in not many millions of years.

A third hypothesis, articulated by a geologist named Ivo Lucchitta, suggests that the lower half of the Grand Canyon might well have been cut by headward erosion within only the past few million years, but that the upper half is much older. That upper half must have been carved (or at least begun) during a time when the Marble Platform itself was overlain with thick layers of Mesozoic rock, from which the river could find a *downhill* angle across the Kaibab Uplift. In this view, the river jumped over the mountain by way of a ramp, but the ramp has since disappeared. The upper layers of rock were stripped away (by some form of surface erosion) from the Marble Platform, leaving that area overshadowed by the Kaibab Uplift. But the uplift by then had a canyon sawn through it.

On the afternoon of Day 8 we beach our boats at Phantom Ranch, one of very few sites within the canyon that connects by steep foot trails with the outside world. The little compound at Phantom includes a campground for hikers from the rims, a set of restrooms, a corral of horses, and a small store. It's the only place where river travelers can buy a glass of cold lemonade, reexperience a flush toilet, and use a pay telephone. The date happens to be September 11, 2001. Mike makes the first call, to his wife, and returns with the day's scarcely believable news.

After a few more calls to loved ones on the outside, we drink our lemonades in silence and then return to the riverbank. We compare what we've heard and pool what we think we know: the World Trade Center leveled, the Pentagon hit, another plane downed near Pittsburgh (or was it Camp David?), perhaps two more hijacked airliners still unaccounted for, 30,000 to 50,000 people dead in

Manhattan, which is being evacuated; the country is shut down, the military are on highest alert, and George Bush is aboard Air Force One, somewhere, headed for Nebraska. Nebraska, I say, that's the underground nuclear command center, Cheyenne Mountain. Cheyenne Mountain is in Colorado, says Margie, who comes from Boulder. You're right, I say. Wait, no, Nebraska is the headquarters of the Strategic Air Command. *What's going on?* we all wonder. I've spoken only with my frail, cheerful, octogenarian parents in Minneapolis, mostly to confirm that they're all right. They are—distraught at the news, like everyone, but not personally assailed by terrorists or sudden turns of ill health. You won't see anything in the sky, says my mother, the planes are all grounded.

We climb into our boats. The rafts pull out, surrendering swoonlike to the current, heading downstream for another ten days in the canyon under conditions of near-total isolation. The other kayakers peel away too, and I find myself alone on the beach. I hesitate. Is there any conceivable reason, I consider, why I should abort this journey and walk out of here? Is there anything useful I can do? Is there anywhere else, right now, I should be? Anywhere else I want to be?

No. I signed on to this trip because I craved an exercise in detachment—from my own life as it has unfolded in recent years, and from the world. So here we are, I think, with an exercise in detachment far more dolorous than I'd foreseen. My sympathies to you, dead and grieving people; good luck, America. I paddle into the heavy current and let it swing me downstream.

We have rapids to run: Horn, Crystal, Serpentine, Bedrock, Upset, and other frivolous challenges to mortality. On the water, we think about the water. In camp, especially when the darkening sky fills with stars and remains peculiarly empty of airplanes, it's different. We think about New York and beyond. We ponder the fact that we're missing a slice of American history, never to be regained, synthesized, or duplicated. We relish unabashedly the simple joys of being together in this marvelous, wild, ancient place.

Me, I'm glad also for the company of Auden. That evening I

reread his poem titled "September 1, 1939." With "clever hopes" expiring at the end of "a low dishonest decade," says the poet,

> *Waves of anger and fear*
> *Circulate over the bright*
> *And darkened lands of the earth,*
> *Obsessing our private lives;*
> *The unmentionable odour of death*
> *Offends the September night.*

It was written, of course, to mark the day Hitler invaded Poland. But the poem is wise beyond old news.

> *I and the public know*
> *What all schoolchildren learn,*
> *Those to whom evil is done*
> *Do evil in return.*

We've camped just above a formidable rapid called Hermit, rated 8. All night, in wakeful moments, we hear its roar. We're now on the threshold of the canyon's more serious water—beginning with Hermit, then Crystal, then a string of other rapids, culminating next week in Lava Falls. At dawn on the morning after our stop at Phantom Ranch, the sky is red. I think: Sailors, take warning.

Several days later, with Crystal safely behind us, we stop to hike up a side canyon called Matkatamiba, a tranquil afternoon's interlude for stretching our legs and gawking at a different sort of scenery. Digressing to explore such byways—with their slots, waterfalls, secret chambers, and polished walls—is an important part of the Grand Canyon experience, a felicity that complements the big-river rush. We've already probed a nice selection: Shinumo Wash, Nautiloid Canyon, Elves Chasm, and the Tapeats Creek trail, which leads to a dramatic waterspout called Thunder River, blasting out of its hole midway down a great Redwall cliff. Matkatamiba Canyon is more graceful than any of them.

We catch a blind eddy at its mouth, leave the boats, and ascend between walls of smoothly curvaceous blue-green limestone, our feet sloshing in clear, warmish water. In some spots the channel, buffed smooth, is only as wide as one human foot. We wade, clamber, and walk several hundred yards before the little canyon bends sharply and, there at its crook, opens out into a natural rotunda. Walls of red rock, hundreds of feet high and undercut with galleries, rise above; the little creek tumbles along its delicate path, across a floor that resembles artfully terraced slate; California redbuds and catclaw acacias, elegantly gnarled like bonsai, stand in patches of rocky soil, and across one spring-moistened slope drapes a profusion of wild grapevines, grasses, and maidenhair ferns, offering a counterpoint texture—cool and green—to all the warm, dry stone. At the center of this extraordinary space is an island of large boulders, like a dais. The whole layout seems to have been designed, perhaps by a subtle Japanese architect, for human ceremony.

Someone says: This would be a great place to hear a concert. Someone else says: This would be a great place to get married. Alluding to a pair of our other kayaking chums, back in Montana, Rick says: "Ron and Carla did get married here."

Married here? It strikes me as an innocent, weird thought from a race of beings to which I don't presently belong. I keep my mouth shut, remembering a December day eighteen years ago, when I myself and a wonderful, serious, joyous woman got married in a beautiful place—on the side of Kitt Peak, with a view of Baboquivari, sacred mountain of the Papago. Ultimately it didn't help.

At the bottom of Upset rapid, stretching wide across the main flow, is a menacing hole. At the top, just beyond the tongue, is a seemingly innocuous diagonal wave, curling off the left wall. The tongue itself isn't glassy and green, not today. Distant rainstorms somewhere upstream have brought a deluge of mocha silt, and the whole river has done a chameleon shift from olive to sullen brown. Even the whitecaps are no longer white. They look like fresh adobe.

After we've scouted the rapid and picked routes for avoiding the hole, Mike takes a leftish line of entry and then, to his shock and ours, finds his raft lifted sideways by the upper wave, which tips him, flips him, as smoothly as a single-blade plow turning dirt. He and Margie, his passenger today, tumble through the air in what seems like choreographed, Hollywood-stunt slow motion. Then they endure the full rapid, dunked through the hole and swept along, trying to catch breaths and get hold of the overturned boat. By the time we reach them, they're in calmer water but still fighting current and cold. Margie, swimming and gasping, grabs hold of my stern handle for a tow to the bank and then, as she climbs out into a jumble of boulders, nearly steps on a rattlesnake, which she hears rattling but can barely see, since her contacts have been splashed ajar. I return to help Rick and John, who are bulldozing the raft toward shore with their kayaks. We get it secured and then, twenty minutes later, with ten pairs of arms lifting and pushing, flipped back upright. The only loss is a lawn chair that wasn't strapped down. Frustrated, embarrassed, Mike says: "I'm gonna take up bowling when I get back to Salt Lake."

The upset at Upset feels like a foreboding prelude. That night over dinner our talk turns with titillating grimness to Lava Falls, which we'll face tomorrow. Bob recalls it vividly from twenty years ago. Steve mentions that the right-side line is difficult at low water, and low water is what we'll have. The right side, Rick says, is *always* a gnarly run. Mike says: My boat needs more ballast; I'll fill the empty carboys with river water. None of the rest of us has ever seen this storied drop. It's a sinister place, with all that lava rock, says Rick. Chase, the thrill-hungry teenager, wants to make multiple runs, riding through on each raft and then running back up to jump aboard the next. Bob asks whether anybody's got a pair of navy-surplus water wings that Chase can use to *swim* the rapid. Don't put that idea in his head, says Cyndi, in her role not as TL but as Mom. You can talk about running Lava this way or that, Steve adds, but you don't really know what you're gonna do until you get there.

Bedding down on the warm sand, I embrace a few resolute thoughts. No point wasting time or energy worrying about things in advance—especially not a mere rapid on a lovely river. If I happen to drown in Lava, which is highly unlikely but possible, it's not important. If I embarrass myself, floundering, swimming for dear life and being rescued, that's even less important. If I manage to slide through with aplomb, less important still. What's important is not to *have done* Lava Falls but to *do* it. What matters is to enter the rapid and live its ten or twenty seconds of magisterial chaos as acutely as possible.

I'm just not sensitive enough, I suppose, to be an angst-ridden person. I sleep soundly, and dream of pretty women and skiing.

We hear it before we see it. Then there's a horizon line, like beveled marble, where the whole river drops away invisibly. Just upstream of the suck, we pull in to scout.

A high cliff of coal-dark basalt looms on the right, a cut-away section of what once was an igneous dam, showing fudgy swirls, puckers, and long rows of columnar basaltic crystals like grinning teeth. We climb. From above on a rocky trail, just the sort of perspective that always makes rapids look deceptively small, Lava looks big. It's not so much a waterfall—despite the name, despite thirty-seven feet of sudden descent—as a raging cascade. Impassable rocks on the left, a hole on the right, a curling wave, another hole, hectic zones of disorderly froth, a big sloping rock at bottom right against which a person would not want to be pinned, and just beside that, another roiling hole, in front of which is a high, tumbling wave. "Busy" is the whitewater term. The right line does, as Rick warned, look uninviting. The left line doesn't exit. There's no sneak route. But there is an imaginable path, from upper right to lower left, nudging past the curlers, crossing a hurricane's eye of relatively calm water, ferrying wide of the lower hole, that each of us commits to mind like a mantra. Then it's back to the boats.

Rick disappears over the horizon line. Bob follows. John signals me from shore: Okay, DQ, your turn. I can see almost nothing as I paddle down the approach tongue. My brain is vacant of any thought more profoundly speculative than *Well, here I go*. As the first waves hit, I hit back, with an aggressive right brace that seems to have been a bit too aggressive, because I find myself in midrapid with my head underwater on the right side. Not wanting to drop entirely upside down (and set up to roll, which would take time), I hold that position for a second or three, hoping that a random upswell might lift me; then, either with such a lift or without it (who knows, who remembers?), I manage to wiggle upright off my very deep brace, finding balance, finding air. I take a few strokes, gather a little momentum, in time to punch my head sideways through the lower wave and miss the hole. As easy as that, I'm in an eddy below, my body aflush with a wave of elated relief.

Rob comments later that I had "an exciting run," which is polite but not complimentary, and that he captured it all on video. TL herself will find irony in the fact that "the most conservative boater had the most exciting run," with which I can't argue. I'm content to know that, perhaps for the first time ever, W. H. Auden (or at least one of his books) has taken a kayak ride through Lava Falls.

Meanwhile we wait vigilantly, Rick and Bob and I, bobbing like flotsam in the left lower eddy, for the others. As Mike's blue raft slides neatly between the lower hole and the sloped rock, he pumps his fist with the joy of redemption. Diz, earnest Diz, forced to run last so that Mark can shoot him in action from *Piglet*'s bow, finds a nifty line, bringing all our remaining Pringles through safely.

On the last evening, our seventeenth on the river, we celebrate with rum punch and begin regretting that the trip has passed so quickly. Like a blink. We're in no great hurry to rejoin the world, however such as the world may now be. We've had almost no news, but within the past few days we've noticed planes reappearing in the sky, evidently on a route between Phoenix and Los Angeles. We can scarcely imagine, or care, what the men and women who sit behind television anchor desks have been saying. Our detachment from the events and aftermath of September 11 has been decreed by circumstance, enforced by isolation, bizarre, cold, not without deep sympathy, and salubrious. The loudest noise down here is the roar of water. Ravens, not newsmen, hover nearby like undertakers. The strata of rock and the silt in the river serve as reminders, thanks to James Hutton, that everything built will be ground down, and that all grinding provides new material for building. At least some of us feel that with enough food, enough river, we could continue indefinitely this mode of travel and life amid this amiable company. But there isn't enough. Size is relative, like time, and in some ways the Grand Canyon is too small. The journey through it is nearly over, already.

As for my personal supplies, I've finished Playfair's *Huttonian*

Theory, but there's still plenty of unread Auden, partly because I've been revisiting favorites. I've joined him repeatedly, for instance, on that evening walk down Bristol Street, past the railway arch, from beneath which comes the voice, claiming:

> *"I'll love you, dear, I'll love you*
> *Till China and Africa meet*
> *And the river jumps over the mountain*
> *And the salmon sing in the street."*

After listening through further such promises of eternal devotion, the eavesdropping poet detects a counterpoint:

> *But all the clocks in the city*
> *Began to whirr and chime:*
> *"O let not Time deceive you,*
> *You cannot conquer Time.*

> *"In the burrows of the Nightmare*
> *Where Justice naked is,*
> *Time watches from the shadow*
> *And coughs when you would kiss."*

My hands ache in the night, pleasantly, from seventeen days of hard use. My shoulders are no worse than when I started. My body has found the river regimen agreeable and my brain has been drawn outside itself. I feel rinsed, peaceful, and whole. I know the end of the poem almost by heart:

> *It was late, late in the evening,*
> *The lovers they were gone;*
> *The clocks had ceased their chiming*
> *And the deep river ran on.*

Next day, our last, we cover six miles of flat water. From the take-out beach at Diamond Creek, as we load our boats into a truck, I can hear the gentle growl of another rapid, just below, waiting to be run.

The Post-Communist Wolf

IT'S TWO HOURS AFTER SUNSET on this snow-clogged Romanian mountain, and in the headlight of a stalled snowmobile stand five worried people and two amused dogs. One of the dogs is a husky. Her name, Yukai, translates from a distant Indian language to mean "Northern Lights." Her pale gray eyes glow coldly, like tiny winter moons. One of the worried people is me. My name translates from Norwegian to mean "cow man" or, less literally, "a cattle jockey who should have stayed in his paddock," neither of which lends me any aura of masterly attunement to present circumstances. The temperature is falling.

Unlike placid Yukai, we five humans are poorly prepared for a night's bivouac in the snow, having long since abandoned most of our gear in an ill-advised gambit to lighten our load and move faster. Three of us—myself, the photographer Gordon Wiltsie, and a German visitor, Uli Geertz—are on backcountry skis with skins, schlepping along steadily behind a biologist named Christoph Promberger and his biologist wife, Barbara Promberger-Fuerpass, who are driving the two snowmobiles. Christoph is a lanky, black-haired German whose almond-thin, lidded eyes make him appear faintly Mongolian—that is, like a young Mongolian basketball player with a wry smile. Though attached to the Munich Wildlife Society, he has worked here in the Carpathian Mountains since 1993, collaborating with a Romanian counterpart named Ovidiu Ionescu, of the Forestry Research and Management Institute, to create the Carpathian Large Carnivore Project. Barbara, a fair-haired Austrian, joined the project more recently and is now beginning a study of lynx. Both of them are hardy souls with considerable field experience in remote parts of the Yukon (where Christoph did a master's degree on wolf biology and where later they honeymooned), so they know a thing or three about winter survival, backcountry travel, problem avoidance, snowmobile repair. But tonight's conditions, reflecting an unusually severe series of January storms and an absence of other human traffic along this road, have caught them by surprise.

Gordon and I are surprised too: that Murphy's Law, though clearly in force, seems unheard of in Romania.

At the outset Christoph was towing a cargo sled, but that had to be cast loose and left behind. Even without it, the Ski-Doos have been foundering in soft six-foot drifts, and much of our energy for the past few hours has gone into pushing these infernal machines, pulling them, kicking them, cursing them, nudging them ever higher toward a peak called Fata lui Ilie; coaxing them and ourselves, that is, ever deeper into trouble. The sensible decision, after we'd bogged at the first steep pitch, then bogged again and again, would have been to turn back at nightfall and retreat to the valley. Instead we went on, convincing ourselves recklessly that the going would get easier farther up. Ha. Somewhere ahead, maybe three miles, maybe five, is a cabin. We have one balky Petzel headlamp, a bit of food, matches, two pairs of snowshoes as well as the skis, but no tent and, since ditching even our packs back at the last steep switchback, no sleeping bags. The good news is that the forest is full of wolves.

"I believe the term is *goat-fucked*," Gordon says suddenly, as though during his last long stretch of silence he's been reading my mind. "A situation that's so absurdly bad, it becomes sublime." Gordon's own situation is more sublime than the rest of ours, since he's suffering from a gut-curdling intestinal flu as well as the generally shared ailments—cold hands, exhaustion, frustration, hunger, and embarrassment. "We could easily spend the night out here, without sleeping bags," he adds.

On that point I'm inclined to disagree: We *could* do it, yes, but it wouldn't be easy.

The purpose of our trip is to reach the Fata lui Ilie cabin and use that as a base for three or four days of wolf-trapping. It's part of the program.

Since 1995, Christoph and his coworkers have collared thirteen wolves, of which five have been shot, two have dispersed beyond the study zone, and four others have fallen cryptically silent, probably when their transmitters failed. One of the missing animals is a female named Timish, the first Carpathian wolf on which Christoph ever laid his hands. Timish, the alpha bitch in a pack, was a savvy survivor, and she opened his eyes to the range of lupine resourcefulness in Romania. Originally trapped and collared in a remote valley near Brasov, a regional capital of some 300,000 people amid the mountains, Timish and her pack soon relocated themselves closer and began making nocturnal forays into the heart of the city. On Brasov's southern fringe was a large meadow where they could hunt rabbits, and by skulking along a sewage channel, then crossing a street or two, they could find their way to a garbage dump, rich with such toothsome possibilities as slaughterhouse scraps, feral cats, and rats. In 1996, Timish denned in the area and produced ten pups. With the aid of a remote camera set fifty meters from the den, Christoph spent many hours watching her perform the intimate chores of motherhood. But times change and idylls fade. Timish disappeared, the fate of her pups is unknown, and in the enterprising ferment of post-Communist Romania, the rabbit-filled meadow is now occupied by a Shell station and a McDonald's.

At the time of our visit, only two wolves are still transmitting, one of which is a male known as Tsiganu, recently collared in another little valley not far from Brasov. Christoph needs more radio-bearing (and therefore trackable) animals for the ongoing study. Hence this night mission to Fata lui Ilie.

The wolf population of the Carpathians is sizable, but the animals are difficult to trap—far more difficult than wolves of the Yukon or Minnesota, Christoph figures—probably because their long history of close but troubled relations with humans has left them warier than North American wolves. Romania is an old country, rich with natural blessings but much wrinkled by conflict and paradox, and history here is a first explanation for everything, including the ecology and behavior of *Canis lupus*. Go back two thousand years, before the imperial Romans put their stamp on the place, and you find the Dacia, a fearsome indigenous people who referred to their warriors as *Daois*, meaning "the young wolves."

Just after World War II, wolves roamed the forests throughout Romania, even the lowland forests, with a total population of perhaps 4,000. They preyed on roe deer, red deer, and wild boar but were much loathed and dreaded for their depredations also against livestock, especially sheep. In the 1950s, the early Communist government, under a leader named Gheorghe Gheorghiu-Dej, sponsored a campaign of hunting, trapping, poisoning, and killing pups at their dens, to reduce the wolf population and make the countryside safe for Marxist-Leninist lambs. That antiwolf pogrom worked well in the lowlands, which were in any case becoming more thoroughly devoted to agriculture and heavy industry. On the high slopes of the Carpathians, though, where lovely beech and oak forests were protected by a tradition of conscientious forestry, where fir and spruce grew to a timberline below tall limestone crags, and where dreams and memories of freedom survived among at least a few of the hardy rural people, wolves survived too.

The Carpathians also served as a refuge for brown bear and lynx. The bear population stands presently at about 5,400, a startling multitude of *Ursus arctos* considering that in all the contiguous western United States (where we call them grizzlies) there are fewer than 1,000. The wolf population, presently numbering somewhere between 2,000 and 3,000 animals, represents a large fraction of all *Canis lupus* surviving between the Atlantic Ocean and Russia. Why has Romania, of all places, remained such a haven for large carnivores? The reasons involve accidents of geology, geography, ecology, politics, and the ironic circumstance that a certain Communist potentate, successor to Gheorghiu-Dej, came to fancy himself a great hunter. This of course was the pipsqueak dictator Nicolae Ceauşescu, who for decades ruled Romania as though he owned it.

Born in the village of Scornicesti and apprenticed to a Bucharest shoemaker at age eleven, Ceauşescu made his way upward as a gofer to early Communist activists during their years of persecution by a fascist regime. He served time in prison, a good place for making criminal and political contacts. He was cunning: he was ambitious and efficacious, though never brilliant; he bided his time, sliding into this opening and then that one, eventually gaining ultimate control as general secretary of the Communist Party in 1965. He styled himself the Conducator, a lofty title that paired him with an earlier supreme leader from Romania's past. He distanced himself from certain Soviet policies, such as the invasion of Czechoslovakia in 1968, and thereby made himself America's favorite Communist autocrat, at least during the administrations of Nixon and Ford and Carter. His manner of domestic governance was merely Stalinism in a Romanian hat, but for a long time the U.S. didn't notice.

Ceauşescu's dark little shadow cast itself across Romania for twenty-five years, with the help of his Securitate apparatus of secret police and informers, which included as many as 3 million people in a nation of just 23 million. "The Securitate maintained a collection of handwriting samples from sixty per cent of the population," according to Robert Cullen, who covered the 1989 revolution for *The New Yorker*. "Anyone with a typewriter had to register it. Mail and telephones were routinely monitored." Such institutional menace wasn't uncommon in the Communist bloc, of course, but it may have weighed more heavily here, owing to a certain wary, fatalistic strain in the national spirit. Romania under Ceauşescu doesn't seem to have had the sort of robust underground network of dissidents that existed in the Soviet Union or, say, Poland or Czechoslovakia. There's a nervous old Romanian proverb, counseling caution: *Vorbesti de lup si lupul e la usa.* Speak of the wolf, and he's at your door.

Ceauşescu's industrial, economic, and social policies were as wrongheaded as they were eccentric. Though he was Stalinist in style, he had that self-important yearning for independence from Moscow, and so he pushed Romania to develop its own capacities in oil refining, mineral smelting, and heavy manufacturing. During the 1970s, his industrialization initiative sucked off a huge fraction of the country's GNP and a big burden in foreign loans; then in the 1980s he became obsessed with paying off those loans and made the Romanian populace endure ferocious austerity in order to do it. He exported petroleum products and food while his own people suffered in underheated and underlit apartments without enough to eat. He instituted a "systematization" campaign, as he called it, which essentially meant bulldozing old neighborhoods and villages in order to force their inhabitants into high-rise urban housing projects, where he could better control their flow of vital resources. His systematization created a larger proletariat living amid ugly urban blight, and his industrialization resulted in some horrendous point-source pollution problems, such as the smelter at Zlatna and the gold-reprocessing plant at Baia Mare, which just recently let slip a vast, wet fart of toxic sludge from one of its containment ponds into the upper Danube drainage, poisoning fish downstream for miles. But for some reason Ceauşescu did *not* become obsessed with exporting timber, and so the Carpathian highlands remained wild and sylvan while other parts of the country grew grim.

The Conducator himself lived a life of splendorous self-indulgence and paranoia, like a neurasthenic king. He had food-tasters to protect him from poisoning. He had germ obsessions like Howard Hughes. He trusted only his wife, Elena, who was his full partner in megalomania and his chief adviser on how to govern poorly. With her, he sealed himself away in palatial residences, letting the people see him mainly through stagy televised ceremonials. For bolstering his ego and political luster he depended also on occasional mass rallies, for which tens of thousands of workers and other citizens were mandatorily mustered to express—or anyway feign—adulation of the Conducator. The last of those, on December 22, 1989, went badly askew and led to his fall. All the other Communist leaders who got dumped during that dizzy time, from Gorbachev down, were content to go peacefully, but Nicolae Ceauşescu required being shot. That speaks not just to the loathsome force of his personality, I suspect, but also to a truth about Romania generally: its edgy, recalcitrant uniqueness.

Ceauşescu's shadow still lingers in some places, including the snowed-over road that may or may not eventually carry us to Fata lui Ilie. The forest is thick. The spruce trees are large and heavily flocked with snow. While the Ski-Doos are mired still again, on another steep switchback below a ridge line, I wonder aloud whether this route was originally cut for hauling timber.

"No, this was a hunting road for Ceauşescu," Christoph tells me. "He'd fly in by helicopter. And his people would come in by four-wheel-drive to organize the hunt." Among other fatuities, Ceauşescu prided himself as a great killer of trophy-size bears. Although his name went into record books and his trophies can still be seen at a museum in the town of Posada, Ceauşescu's actual accomplishments were contemptible: squeezing off shots at animals that had been located, fattened, and baited for his convenience. The sad irony is that so long as he arrogated the country's bear-hunting rights largely to himself and allowed his forestry bureaucrats to protect the habitat, the bear population flourished. Records show that it peaked, at about eight thousand animals, in 1989. The end of that year was when the ground shifted for everyone—carnivores, citizens, and Ceauşescu. "Until December," Robert Cullen noted, "the vast majority of the Romanian people feared the Securitate and submitted wearily to its control." Then, on December 22, the people arose and Ceauşescu, losing his nerve, tried to flee but was captured. On Christmas Day, before a firing squad, the great hunter got his.

Farther along, when we pass a spur road to Ceauşescu's helicopter pad, I feel tempted to ski up and inspect it. But by now Christoph and Barbara are far ahead on the snowmobiles, Gordon is with them, and I'm skiing through darkness with only Uli's dim headlamp as a point of guidance. Ceauşescu is dead, the bears are asleep, the new government is led by a center-right coalition of parliamentarians, the Carpathian forests are being privatized to their great peril, the currency is weak, the mafia is getting strong, wolves are what brought me up onto this mountain, and all idle contemplation of the pungent contingencies of recent Romanian history is best left, I realize, for a time when I'm not threatened by hypothermia.

The wolf known as Tsiganu was trapped on December 19, 1999, near a valley called Tsiganesti. The handling, collaring, and release were done by a Romanian technician named Marius Scurtu, a sturdy young man with an unassuming grin and a missing front tooth, from Ovidiu Ionescu's wildlife unit at the forestry institute. Marius had blossomed into an important member of the carnivore project, absorbing well the field training in wolf capture that Christoph gave him and showing great appetite for the hard back-country legwork. In recognition of his role, he was allowed to christen the new animal. Besides relating the wolf to that particular valley, the name he picked—Tsiganu—means "Gypsy."

At the time of trapping, Tsiganu weighed ninety-five pounds. He was notable for the lankiness of his legs and, after careful measurement, the length of his canine teeth. Since collaring, he has rejoined a small pack of four or five animals, though whether he himself is the alpha male remains uncertain. He now broadcasts his locator beeps on a frequency of 148.6 megahertz, and several times each week either Marius or another project technician goes out with a map, a radio receiver, and a directional antenna to check on him. Tsiganu seldom lets himself be seen, but from his prints and other evidence in the snow, a good tracker can learn what he has been doing. In the past month he has killed at least three roe deer, two dogs, and two sheep. On a warmish day not long before our misadventure on the trail toward Fata lui Ilie, Gordon and I skied along with a tracker named Peter Suerth.

We followed Peter up a tight little canyon into the foothills above a village. It was slow travel, through wet heavy snow along the bank of a small stream, but within less than a mile we came to a kill. The rib cage and hide of a roe deer, partly covered by overnight snowfall, confirmed that Tsiganu and his pack hadn't gone hungry. Continuing upward, we passed an old log barn within which, by their companionable gurgles and their neck bells, we could hear sheep, safely shut away behind a door. Moments later we met a man in country clothes, presumably the sheep-owner, trudging down a steep slope. Peter spoke a few words with him, then told us the gist of the exchange. *Wolves, you want wolves?* the man had said. *Wolves we've got, around here. Lots of them.*

We angled steeply up a slope, rising away from the creek bottom. A half-hour of climbing brought us each to a full sweat, and onto a ridge. Peter took another listen with the receiver, catching a strong signal that seemed to place Tsiganu within 300 yards. Which direction? Well, probably *there*, to the northwest. But the tempo of beeps also indicated that the animal was active, not resting, and therefore his position could change fast. We hustled northwest along the ridge line. When Peter listened again he got a very different bearing, this one suggesting that Tsiganu and his pack were below us, possibly far below, on the opposite slope of the creek valley we'd just left. Or maybe the earlier signal had been deceptive, because of echo effects from the terrain. Or maybe this one was the echo.

While Peter pondered those uncertainties, I noticed that we had skied our way up to the southeastern outskirts of a place I recognized—a snowbound hamlet of thatch-roofed cottages, conical haystacks, coppiced willow stumps defining an unplowed lane, and a few shapely farmhouses with gabled and turreted tin roofs, all hung like a saddle blanket across the steep sides of this foothill ridge. It was called Magura. It seemed a mirage of bucolic tranquillity from the late Middle Ages, but it was real. I had been here before.

Most recently, I had been here with a Romanian friend, Andrei Blumer, when he and Gordon and I skied up from the other side, on a day of bright sunshine and stabbing cold, and stopped to visit an elderly couple named Gheorghe and Aurica Surdu. The Surdus live in a trim little cottage they built fifty years ago to replace a five-hundred-year-old cottage on the same spot, in which Aurica had been born. Aurica is a pretty woman of seventy-some years, with a deeply lined face and a wide, jokey smile. We were greeted effusively by her, Gheorghe, and their middle-aged son, another Gheorghe but nicknamed Mosorel, who himself had boot-kicked up through the snow for a Saturday visit. Passing from deep snowbanks and icy air into a small narrow room with a low ceiling, a bare bulb, and a woodstove upon which simmered a pot of rose-hip tea, we commenced to be steam-cooked with hospitality. Aurica, wearing a head scarf and a thick-waled corduroy vest, spoke as little

English as Gordon and I did Romanian, but she made herself understood, and her motherly eyes missed nothing. She stood by the stove and fussed cheerily while Andrei traded news with Mosorel, Gordon thawed his camera lenses, and I waited for my glasses to clear. *Have some rose tea, you boys, get warm. Here, have some bread, have some cheese, don't be so skinny.* Okay, thanks, don't mind if we do. The tea was deep-simmered and laced with honey. *Have some smoked pork. And the sausage too, it's good, here, I'll cut you a bigger piece, don't you like it? You do? Then don't be shy, eat. Have some of the apple.* We had set off without lunch, so we were pushovers. *Mosorel, give them some tsuica, what are you waiting for?* Mosorel, grinning broadly, poured us heated shots of his mother's homemade apple-pear brandy, lightly enhanced with sugar and pepper. Tsuica is more than just the national moonshine; it's a form of communion, and we communed.

Mosorel's right hand was swaddled in a large white bandage. It testified to a saw accident several months earlier, Andrei explained, in which Mosorel had sliced off his pinkie and broken his fourth finger while cutting up an old chest for usable lumber. Mosorel is a carpenter, sometimes. Sometimes too he's a tailor; his nickname means, roughly, "Mr. Thread." Until the saw accident, he had also been pulling shifts at a factory down in the nearby town. Like his parents, who still raise pigs, cows, sheep, onions, corn, beets, potatoes, and more than enough apples and pears for tsuica, Mosorel is a versatile man of diverse outputs. The hand injury didn't seem to damper his spirits, possibly because some joyous aptitude for survival runs like a dominant gene through the family, homozygous on both sides of his parentage. As the sweet liquor spread its heat in our bellies, the talk turned in that direction—to survival, and how its terms of demand had changed.

During the Communist era, Gheorghe and Aurica Surdu had been required to supply eight hundred liters of milk each year to the state. Andrei translated this fact, Aurica nodding forcefully: *Yes, eight hundred.* There were also quotas to be met in lambs, calves, and wool. Since the revolution, things had changed; no longer are Gheorghe and Aurica obliged to deliver up a large share of their farm produce, but market prices are so low that rather than selling their milk, they feed it to the pigs. So, I asked simple-mindedly, is life better or worse since the fall of Ceauşescu? The talk rattled forward in Romanian for a few moments until Andrei paused, turned aside, and told me that Mosorel had just said something important.

"At least we're not scared now," he had said.

Just below the high village of Magura, at the mouth of the small river valley draining from Fata lui Ilie and other peaks, sits a peculiar little town called Zarnesti. Narrow streets, paved with packed snow at this time of year, run between old-style Transylvanian row houses tucked behind tall courtyard walls closed with big wooden gates. Horse-drawn sleighs jingle by, carrying passengers on the occasional Sunday outing. Heavy horsecarts with rubber tires haul sacks of corn, piles of fodder, and other freight. Young mothers pull toddlers and grocery bags on little metal-frame sleds. Kids ice-skate down the glassy snow-packed lanes. There are also a few automobiles—mostly beat-up Romanian Dacias—creeping between the snowbanks, and along the southern edge of town rises with sudden ugliness a cluster of five-story concrete apartment blocs from the Communist era, like a histogram charting the grim triumph of central planning. Beside the train tracks sits a large pulp mill that eats trees from the surrounding forests, digests them, and extrudes the result as paper and industrial cellulose. The mill site is cluttered with cranes, tanks, conveyors, piles of logs, a long eyeless building stuccoed in weary pink, and a few smoke-stacks. Beyond it is another neighborhood of concrete high-rises.

You can walk all afternoon among the winding lanes of Zarnesti, down to the main street, past the Orthodox church, past the pulp mill, looping back through the post office square, and not see a single neon sign. There are no restaurants and no hotels—none that I've managed to spot, anyway. Yet the population is 27,000. People live here and work here, but few visit. For decades Zarnesti was a closed town. The reason for its closure was security strictures related to the other industrial plant, over near the police station, the one commonly known as "the bicycle factory." The bicycle factory was really a munitions factory, founded in 1938, when Romania was menaced by bellicose neighbors during the buildup toward World War II; later, in the Communist era, it had thrived and diversified. It produced artillery, mortars, rockets, treads for heavy equipment, boxcars, and—yes, as window dressing—a few Victoria bicycles. This is the factory where Mosorel worked until mutilating his hand. For decades it was Zarnesti's leading industry. But the market for Romanian-made rockets and mortars has been wan since the disintegration of the Warsaw Pact, and the bicycle factory, which once employed 13,000 people, has laid off about 5,000 since 1989. At the pulp mill, likewise, the workforce has shrunk to a fraction of its former size. The town's economy—at least the old economy, fed by geopolitical suppositions and mandates that flowed up from Bucharest—now resembles a comatose patient on a gurney, ready to be wheeled who knows where. Still, Zarnesti is filled with stalwart people, and a few of those people are energized with new ideas and new hopes.

One new idea is large-carnivore ecotourism. It began in 1995, when Christoph Promberger was contacted by a British group who had heard about the Carpathian Large Carnivore Project and wanted to bring paying visitors to this remote corner of Europe for a chance to see wolves and bears. They came—not actually to Zarnesti but to another small community nearby—and the money spent on lodging and food, though modest, was significant to the local economy. Two years later Christoph and his colleagues repeated the experiment as an independent venture. They welcomed eight different tour groups totaling some seventy people, who were accommodated in small *pensiunes,* vacation boarding-houses run by local families. Although the likelihood of actually glimpsing a wolf or a brown bear in the wild is always low, even for experienced trackers like Marius and Peter, some nature-loving travelers were quite satisfied, it seemed, to hike or ride horses through Carpathian forests in which a sighting, or a set of tracks, was always possible. Meanwhile the wolf fieldwork came to focus on the wooded foothills and flats of the Barsa Valley, which stretches thirty miles into the mountains above Zarnesti. And adjacent to the Barsa is a newly enlarged protected area called Piatra Craiului Natural Park, a massif of limestone crags, high forests, and alpine meadows harboring several endemic plant species. Piatra Craiului, now supported with a grant from the Global Environmental Facility of the World Bank, has its own great potential as a tourist destination but little such traffic so far. Christoph discussed the tourism opportunity with a couple of venturesome folks in the town. Large carnivores, he pointed out, might attract travelers who wouldn't come just for edelweiss and primroses. One man he talked to was Gigi Popa.

Gigi Popa is a forty-six-year-old businessman whose trim mustache, balding crown, and gently solicitous manner conceal the soul of a risk-taker and a performer. Give him three shots of tsuica, a guitar, and an audience—he'll smile shyly, then hold the floor for an evening. Give him a window of economic opportunity—he'll climb through it. Gigi grew up in a small village near Zarnesti, the son of a sheetmetal worker. In the 1980s, he worked as a cash-register repairman for a large, inefficient government enterprise charged with servicing machines all over Romania. The machines in question were mediocre at best, and destined to be obsoletized by modern electronic versions. Gigi couldn't divine all the coming upheavals, but he could see clearly enough that mechanical Romanian cash registers were not a wave to ride into the future.

He and his wife lived in a little house behind a high courtyard gate, a place that was charming and solid but had lately come under threat to be leveled for more concrete apartments, in accordance with President Ceauşescu's systematization campaign. As the campaign approached to within wrecking-ball distance, Gigi and his family could do nothing but watch and dread. Then, blessed surprise, Nicolae Ceauşescu himself fell before Gigi's house did.

"After the revolution, I change quickly my job and my direction," Gigi says. He got out of cash-register repair and opened a small store in the back of the house.

He was ready for the next step, not knowing what the next step might be, when Christoph told him about English, Swiss, and German travelers who would be coming to Zarnesti, drawn by the wolves in the mountains but needing lodging in town. Gigi promptly remodeled his home and his identity again. He became a *pensiune* keeper, with four guest rooms ready the first summer and another four the following year. He now plays an important partnership role to the Carpathian Large Carnivore Project's program of tourism. Gigi's *pensiune* is where Gordon and I have been sleeping, for instance, when we're not sublimely geschtuck in the mountains.

One morning I ask Gigi the same question I asked Mosorel: Has the new order made life better or worse? "The good thing of the revolution is everybody can do what he have dreams," Gigi says. "Because *everybody* have dreams. And in Ceauşescu time you can do no thing for your own. Must be on the same . . . same . . ."—he makes a glass-ceiling gesture—". . . *level*. Everybody." Whereas now, he says, a person with initiative, wit, a few good ideas, and a willingness to gamble on them can raise himself and his family above the dreary old limit. The bad thing, he says, is that free-market entrepreneurship involves far more personal stress than a government job in cash-register maintenance.

One day in the summer of 1999, Christoph and Barbara noticed a sizable construction job under way in the Barsa Valley, some miles upstream from Zarnesti. The foundation was being laid for a hundred-room hotel.

This was not long after Christoph had begun discussions with Gigi Popa and a few other local businessmen, as well as the town mayor, about not just tourism but a vision of sustainable ecotourism for Zarnesti. The crucial premise of that vision was to let the Barsa Valley remain undeveloped—and thereby to preserve an intact riparian ecosystem, as well as habitat for large carnivores, with their attractive appeal to foreign visitors—while the infrastructure to support those visitors would be built as small-scale operations down in the town. If the valley itself was consumed by suburban sprawl and recreational development, Christoph had explained, then the carnivore habitat would be badly fragmented, if not destroyed, and the Large Carnivore Project would be forced to move, taking not just its research focus but also its ecotourism activities elsewhere. But if the Barsa habitat was protected, then the project could remain, channeling visitors to whatever small *pensiunes* might be available in Zarnesti. Everyone had seemed to agree that this was the sensible approach. Yet now the hotel construction revealed that someone else—an investor from the city of Brasov, fifty miles away—intended to exploit the proximity of Piatra Craiului Natural Park on an ambitious scale. And belatedly it was revealed that the town council had approved open development zoning for the entire valley. "So this was disaster," Christoph remembers thinking. "Absolute disaster."

Christoph himself had to leave the country just then, for a short visit back in Germany. Fortunately, Andrei Blumer had by that time joined the project as a specialist in rural development. Together they shaped their best argument for valley protection plus in-town entrepreneurship, so that Andrei could present that argument to the mayor.

Zarnesti's mayor is a mid-fortyish man named Gheorghe Lupu, formerly an engineer in the bicycle factory before Romanian bicycles lost their tactical military appeal. Bright and unpretentious, his dark hair beginning to go gray, Mr. Lupu wears a black leather jacket at work, keeps his office door open to drop-by callers, and describes himself jokingly as a "cowboy mayor." About the problems of Zarnesti, though, he's serious. Tax revenues yield only 10 percent of what they did before the revolution; the pulp mill has laid off two thousand people, the bicycle factory even more; the sewage system and the gas supply network need work; the roads too cry out for repair. There was little basis to assume that this harried man would muster much sympathy for protecting wolf habitat, notwithstanding the fact that his own name, Lupu, translates as "wolf." But would he be able at least to grasp the connection between large carnivores, open landscape, and tourism? It was a tense juncture for Christoph, having to absent himself while the whole Barsa Valley stood in jeopardy.

Just before leaving for Germany, he received a terse electronic message on his mobile phone. It was from Andrei, saying: "Lupu stopped everything." The mayor had moved to reverse the council's decision. Let the tourists eat and sleep in Zarnesti, he agreed, and pay their visits to the wild landscape as day-trippers. He had embraced the idea of zoning protection for the valley.

But to announce a policy of protection is one thing; real safety against the forces of change is another. Barbara and I get a noisy reminder of that difference during an excursion to set traps for her lynx study.

We're twenty-some miles above Zarnesti, where the Barsa road narrows to a single snowmobile trail. Barbara has driven her Ski-Doo, loaded with custom-made leg-hold traps and other gear, me riding my skis at the end of a tow rope behind. In the fresh snow at trailside we've seen multiple sets of lynx prints as well as varied signs of other animals—deep tracks from several red deer that came wallowing down off a slope, fox tracks, even one set from a restless bear that has interrupted its hibernation for a stroll. Late in the afternoon, just as Barbara finishes camouflaging her last trap, we hear the yowl of another snowmobile ascending the valley. At first I assume that it must be Christoph's. But as the machine throttles back, I see it's a large recreational Polaris driven by a middle-aged stranger in a fur hat, with a woman on the seat behind him. Then I notice that Barbara has stiffened.

She exchanges a few sentences in Romanian with the stranger. He seems rather jovial; Barbara speaks curtly. The man swings his snowmobile around us and goes ripping on up the valley. When he's beyond earshot, which is instantly, Barbara explains what just transpired.

Claims he's from Brasov, she says. But he is *not* Romanian, to judge from his accent. Probably a wealthy Italian with a second home. When he heard what Barbara was doing—setting traps to catch lynx—he thought she meant trapping for pelts, and he acted snooty; when she added that it's for a radio-tracking study, he graced her with his patronizing and ignorant approval. *Oh, you're doing wildlife research—okay.* His ladyfriend, on the other hand, was worried. "She asked if it would be dangerous to continue, with all the lynx in here. *Ya*, it would," Barbara says caustically. "Keep out." The upper valley is closed to joy-riding traffic and those two have no business being here, Barbara explains. Unlimited motorized access, along with development sprawl and other symptoms of the new liberty and affluence, are now a damn sight more threatening to the lynx population—and the wolves, and the bears—than fur-trapping, judicious timbering, or even the crude, spoliatory hunting once practiced by Nicolae Ceauşescu, with all his minions and helicopter pads.

Barbara has never before seen a recreational snowmobile in Zarnesti, let alone up here. "Aaagh," she says, as the roar of the Polaris fades above us. "It all starts with one. There are *so* many rich guys in Brasov now."

On the following day, the last Sunday of January, the pattern of nightly snowfalls and frigid temperatures breaks to a thaw. Down in Zarnesti, the lovely deep drifts on roof eaves begin to sag weightily. By noontime the main streets are full of slush. The polished white lanes near Gigi Popa's *pensiune*, where the horse sleighs and skating children lent such a flavor of timeless grace, have turned to mush. Cars flounder like mud-bound elephants. From being cold, hard, and gorgeous, the weather has turned warm and ugly. We have survived our reckless excursion to Fata lui Ilie, descended from that mountain, and now the valley is showing us a very different sort of face.

Gordon departs for America, and Uli Geertz heads home to Hamburg. I stay behind. In early afternoon I set off in my rental car through a sleety drizzle toward the city of Brasov. The rain makes everything seem grimier and more joyless, especially the half-idle pulp mill at the edge of town. Just beyond that, I find myself stopped at a railroad crossing. In my fog of rumination, thinking about the death of the old regime and the rise of the new one, about carnivore research and ecotourism, about all the odds stacked against the possibility that big predators will survive on our planet much longer, I'm slow to recognize that the vehicle in front of me is Christoph's. He climbs out and comes walking back.

"Marius just called," Christoph says. "Tsiganu's been shot. He may be dead."

The details are still blurry, but it seems that a couple of boar hunters let fly at the wolf for no particular reason except his wolfhood. Probably they were poaching, since no gamekeeper was present, as mandated for a legitimate boar hunt. Tsiganu is wounded, hard to say how badly, but still on his feet at last report. Marius heard the shots. He came upon the hunters a few moments later. Marius is still out there, Christoph tells me, following a trail of radio beeps and blood spoor through the wet snow. Before long he will either find Tsiganu's fresh carcass or else run out of daylight without knowing quite what's what.

Having told me this much, Christoph jumps back in his car and the train barrier lifts.

I do my business in Brasov, distractedly, and return to Zarnesti. A day passes. Still there's no definite news of Tsiganu. On the morning of the second day, again a warm one, I set out tracking with Marius and two project assistants.

We park the Dacia truck on a roadside above a village and begin hoofing along a farm lane into the foothills. At first we slog through slush and mud, then up into knee-deep snow, then still higher into a zone where the crust has barely softened. We follow a snowed-over lane on a climbing traverse between meadows, along wooded gullies, beyond the last of the farmhouses, the last of the barking dogs, past two men hauling logs with a pair of oxen. Marius moves briskly. He's a short, solid fellow with good wind and a long stride. He cares about this animal—both about *Canis lupus* as a denizen of the Romanian mountains, that is, and about Tsiganu as an individual. But Marius is a home-bred Romanian forestry worker, not a foreign-trained biologist, and his attitude is complexly grounded in local realities.

"Last year the wolf was killing for me two sheep," he says as we walk. "Because the shepherd was drunk. Was like an invitation to eat." The shepherd was his employee, helping Marius raise a few animals on the side. Some farmers moan about such losses, but what do they expect? Marius wonders. That the wolf, which has lived as a predator in these mountains for thousands of years, should now transform itself into a vegetarian? As for hunters who would offhandedly kill a wolf for its fur, he can't comprehend them. "Also I am a hunter," he says. He shoots ducks, pheasants, wild boar, and in self-defense he wouldn't hesitate to kill a bear. But a wolf, no, never. It's much nicer simply to go out with his dogs, hike in the forest, and know that in this place the ancient animals are still present.

Two miles in, we pick up a signal from Tsiganu's collar. The bearing is south-southwest, toward a steep wooded valley that descends from a castle-shaped rock formation among the peaks above. Farther along, we get another signal on roughly the same line, and now the tempo of beeps indicates that Tsiganu is alive— at least barely alive, because he's moving. Here we split into two groups, for a better chance of cutting his trail. Marius and I continue the traverse until we cross a single set of wolf tracks, then back-follow them up a slope. The tracks are deep, softened in outline by at least one afternoon's melting, and show no sign of blood. Yesterday? Or earlier, before the shooting? They might be Tsiganu's or not. If his, is the stride normal? Has his wound already clotted? Or is he lying near death with a slug lodged against his backbone, or in his lung, or in his jaw, while his packmates have gone on without him? Are these in fact his tracks, or some other wolf's? No way of knowing.

So we hike again toward the radio signal, post-holing our way through knee-deep crust. We round a bend that brings us face-on to the valley below the castle-shaped peak. Here the radio signal gets stronger. We stare upward, scanning for movement. We see none.

Marius disconnects the directional antenna from the receiver. He listens again, using the antenna cable's nub like a stethoscope, trying to fine-focus the bearing. Again a strong signal. So we're close now. Maybe one hundred meters, Marius says. He tips back his head and offers a loud wolfish howl, a rather good imitation of a pack's contact call. We listen for response. There's a distant, dim echo of his voice coming off the mountain, followed by silence. We wait. Nothing. We turn away. I begin to fumble with my binoculars.

Then from up in the beeches comes a new sound. It's Tsiganu, the Gypsy carnivore, the post-Communist wolf, howling back.

The Megatransect

AT 11:22 ON THE MORNING OF SEPTEMBER 20, 1999, J. Michael Fay strode away from a small outpost and into the forest, in a remote northern zone of the Republic of Congo, setting off on a long and peculiarly ambitious hike. By his side was an aging Pygmy named Ndokanda, a companion to Fay from adventures past, armed now with a new machete and dubiously blessed with the honor of cutting trail. Nine other Pygmies marched after them, carrying dry bags of gear and food. Interspersed among that troop came still other folk—a camp boss and cook, various assistants, Michael ("Nick") Nichols with his cameras, and me.

It was a hectic departure to what would eventually, weeks and months later, seem a quiet, solitary journey. Fay planned to walk across Central Africa, more than a thousand miles, possibly much more, on a carefully chosen route through untamed regions of rainforest and swamp, from northeastern Congo to the coast of Gabon. It would take him at least a year. He would receive resupply drops along the way, communicate as needed by satellite phone, and rest when necessary, but his plan was to stay *out there* the whole time, covering the full route in a single uninterrupted push. He would cross a northern stretch of the Congo River basin, then top over a divide and descend another major drainage, the Ogooué.

Any big enterprise needs a name, and Fay had chosen to call his the Megatransect—*transect* as in cutting a line, *mega* as in mega, a label that variously struck those in the know as amusing or (because survey transects in field biology are generally straight and involve statistically rigorous repetition) inappropriate. Fay is no sobersides, but amusement was not his intent. Behind this mad lark lay a serious purpose—to observe, to count, to measure, and from those observations and numbers to construct a portrait of great Central African forests before their greatness succumbs to the inexorable nibble of humanity. The measuring began now. One of Fay's entourage, a bright young Congolese named Yves Constant Madzou, paused at the trailhead to tie the loose end of a string to a small tree.

I paused beside him, because I'd heard about the string and it intrigued me. In the technical lingo, it was a *topofil*. Its other end was wound on a conical spool inside a Fieldranger 6500, a device used by foresters for measuring distance along any walked route. The topofil pays out behind a walker while the machine counts traversed footage, much as a car's odometer counts traversed miles. Each spool holds a six-kilometer length. Madzou carried a half-dozen extras, and somewhere among the expedition supplies were many more. Being biodegradable, the string would quickly disappear down the gullets of termites and other jungle digesters, I'd been told, but the numbers it delivered with such Hansel-and-Gretel simplicity would be accurate to the nearest twelve inches. You can't get that precision from a Global Positioning System (GPS) and a map. Running the topofil each day from the red plastic box on his belt was to be one of Madzou's assignments.

Now, as he stepped out after Fay in the first minutes of Day 1, the Fieldranger gurgled in a low, wheezy tone, like an asthmatic retriever catching its breath between ducks. Madzou trailed filament like a spider. The string hovered, chest-high, under tension. And I found it pungent to contemplate that if Fay's expedition proceeds to its fulfillment, a thousand-mile length of string will go furling out through the equatorial jungle. That string seemed an emblem of all the oxymoronic combinations this enterprise embodies—high tech and low tech, vast scales and tiny ones, hardheaded calculation and loony daring, strength and fragility, glorious tropical wilderness and a mitigated smidgen of litter. As he walks, Fay will gather data in many dimensions by many means, including digital video camera, digital audio recorder, digital still camera, notebook and pencil, GPS, conductivity meter, thermohygrometer, handheld computer, digital caliper, and hand lens. The topofil will be a quaint but important complement to the rest.

Within less than an hour on the first day we're shin-deep in mud, crossing the mucky perimeter of a creek. "Doesn't take long for the swamps to kick in around here," Fay says cheerily. He's wearing his usual outfit for a jungle hike: river sandals, river shorts, a lightweight synthetic T-shirt that can be rinsed out each evening and worn again next day, and the day after, and every day after that until it disintegrates. River sandals are preferable to running shoes or tall rubber boots, he has found, because the forest terrain of northeastern Congo is flat and sumpy, its patches of solid ground interlaced with leaf-clotted spring seeps and blackwater creeks, each of them guarded by a corona of swamp. A determined traveler on a compass-line march is often obliged to wallow through sucking gumbo, cross a waist-deep channel of whiskey-dark water flowing gently over a bottom of white sand, wallow out through the muck zone on the far side, rinse off, and keep walking. Less determined travelers, in their Wellingtons and bush pants, just don't get to the places where Fay goes.

He stops to enter a datum into his yellow Rite-in-the-Rain notebook: elephant dung, fresh. Blue-and-black swallowtail butterflies flash in sun shafts that penetrate the canopy. He notes some fallen fruits of the plant *Vitex grandifolia*. Trained as a botanist before he shifted focus to do his doctorate on western lowland gorillas, Fay has an impressive command of the botanical diversity on which big mammals depend—he seems familiar with every tree, vine, and herb. He knows the feeding habits of the forest elephant (*Loxodonta cyclotis*, the smaller species of African elephant adapted to the woods and soggy clearings of the Congo basin) and the life cycles of the plants that produce the fruits it prefers. He can recognize, from stringy fecal evidence, when a chimpanzee has been eating rubber sap. He can identify an ambiguous tree by the smell of its inner bark. He sees the forest in its particulars and its connectedness. Now he bends pensively over a glob of civet shit. Then he makes another notation.

"Mmm. This is gonna be fun," he says, and walks on.

Mike Fay isn't the first half-crazed white man to set out trekking across the Congo basin. In a tradition that includes such Victorian-era explorers as David Livingstone, Verney Lovett Cameron, Savorgnan de Brazza, and Henry Morton Stanley, he's merely the latest. Like Stanley and some of the others, he has a certain perverse gift for command, a level of personal force and psychological savvy that allows him to push a squad of men forward through difficult circumstances using a mix of inspirational goading, promised payment, sarcasm, imperiousness, threat, tactical sulking, and strong example. He's a paradoxical fellow and therefore hard to ignore, a postmodern redneck who chews Red Man tobacco, disdains political correctness, knows a bit about tractor repair and a lot about software, and views the crowded, suburbanized landscape of modern America with cold loathing. Born in New Jersey, raised there and in Pasadena, he sees no going back; he'll live out his life and die in Africa, he says. What makes him different from those legendary Victorian zealots is that he's not traveling in service of God or empire or the personal enrichment of the king of Belgium. He does have sponsors, most notably the National Geographic Society, and also the Wildlife Conservation Society (WCS) of New York, for which he's a staff member on paid leave, but he's certainly not laboring for the greater glory of them. His driving motive—or rather, the first and most public of his *two* driving motives—is conservation.

His immediate goal is to collect a huge body of diverse but intermeshed information about the biological richness of the ecosystems he'll walk through and about the degree of human presence and human impact. He'll gather field notes on the abundance and freshness of elephant dung, leopard tracks, chimpanzee nests, and magisterial old-growth trees. He'll make recordings of birdsong for later identification by experts. He'll register precise longitude-latitude readings every twenty seconds throughout the walking day, with his Garmin GPS unit and the antenna duct-taped into his hat. He'll collect rock samples, note soil types, listen for half a dozen different species of skrawking monkey. He'll detect gorillas by smell and by the stems of freshly chewed *Haumania dankelmaniana*, a monocot vine they munch like celery. Beyond the immediate goal, his ultimate purpose is to systematize those data into an informational resource unlike any ever before assembled on such a scale—and to see that resource used wisely by the managers and the politicians who will make decisions about the fate of African landscapes. "It's not a scientific endeavor, this project," Fay acknowledges during one of our talks before departure. Nor is it a publicity stunt, he argues, answering an accusation that's been raised. What he means to do, he explains, is to "quantify a stroll through the woods."

Then there's his second driving motive. He doesn't voice it explicitly, but I will: Mike Fay is an untamable man who just loves to walk in the wilds.

Completing this marathon trek won't be easy, not even for him. There are dire diseases, minor health hassles, political disruptions (such as the civil war that racked the Republic of Congo in 1997), and other mishaps that could stop him. He's familiar with malaria, aware of filariasis and Ebola, and has found himself inconveniently susceptible to footworms, a form of parasite that can travel from elephant dung into exposed human feet, burrowing tunnels in a person's toes, only to die there and fester. He's aware that every scratch on an ankle or an arm in this feculent environment is a potential infection. He has tasted the giddy vulnerability of facing armed poachers unarmed, confiscating their meat, burning their huts, and wondering bemusedly why they didn't just kill him. But the biggest challenge for Fay will come *after* all his walking.

Can he make good on the claim that this encyclopedia of field data will be useful? Can he satisfy the doubters that it isn't just a stunt? Can he channel his personal odyssey into practical results for the conservation of African forests?

He's very stubborn; maybe he can.

Suddenly, two kilometers on, Fay makes a vehement hand signal: *stop*. As we stand immobile and hushed, a young male elephant appears, walking straight toward us through the understory.

Ndokanda slides prudently to the back of the file, knowing well that a forest elephant, nearsighted and excitable, is far more dangerous than, say, a hungry leopard or a runaway truck. Fay raises the video camera. The elephant, visually oblivious and upwind of our smell, keeps coming. The videotape rolls quietly. When the animal is just five yards from him and barely twice that from the rest of us, too close for anyone's comfort, Fay says in a calm voice: "Hello." The elephant spooks, whirls around, disappears with its ears flapping.

Tusk length, about forty centimeters, Fay says. Maybe ten or twelve years old, he estimates. It goes into his notebook.

Mike Fay is a compact forty-three-year-old American with a sharp chin and a lean, wobbly nose. Behind his wire-rimmed glasses, with their round, smoky lenses, he bears a disquieting resemblance to the young Roman Polanski. Say something that's doltish or disagreeable and he'll gaze at you silently the way a heron gazes at a fish. But on the trail he's good company, a man of humor and generous intellect. He sets a punishing pace, starting at daylight, never stopping for lunch or rest, but when there are field data to record in his yellow notebook, fortunately, he pauses often.

He first came to Central Africa in 1980, after a stint with the Smithsonian Peace Corps (a scientific variant of the U.S. Peace Corps) doing botany in Tunisia. He signed up for another stint on the understanding that he'd go to a new national park in the Central African Republic, near its borders with Chad and Sudan. The park, known as Manovo-Gounda St. Floris, was then just wishful lines on a map. The lines encircled an area rich with wildlife, in a region over which the CAR government exerted virtually no control. It was a savanna ecosystem, fertile and wild, supporting large populations of elephant, black rhino, giant eland, kudu, giraffe, roan antelope, and other big mammals. "A million hectares," Fay tells me, "and you're the only white man in those million hectares for eight months out of the year. It was like paradise on Earth." Yet it wasn't so paradisaical when Chadian and Sudanese poachers came to slaughter the elephants. Both his love for Central Africa and his ferocity as a conservationist seem to be rooted in that place and time.

It was at St. Floris too that Fay began to—what's the right phrase? *go AWOL? step off the ranch? disappear into nowhere for long periods?*—let's say *leaven* his more focused scientific work with wildcat exploratory journeys. Since the park's landscape was open and flat, he put his Peace Corps–issue Suzuki 125 trail bike to some unauthorized use. "I decided that the way to really see that place was to take long traverses from one road to another, sometimes seventy or eighty kilometers, across the places where no one had ever been." Too many field biologists, in his judgment, never venture more than a few kilometers from their base camps. Fay rejected such tethering; he hungered to see the wider scope and the interstitial details. He was restless. He would load the little bike with extra fuel, a patch kit for flats, two weeks' worth of food, and go.

We leave camp just after dawn on Day 3 and follow the Mopo River downstream along a network of elephant trails. We're a smaller group now, Nick Nichols and his assistant having backtracked to the start for other work, intending to rendezvous with Fay's march some weeks later. Fay, Madzou, and I set out while the crew are still eating breakfast, giving us a relatively quiet first look at forest activity. Under a high canopy of *Gilbertiodendron* trees, the walking is easy. The understory is sparse, as it generally tends to be in these dominant stands of *Gilbertiodendron*, and well trampled by elephant traffic. Later, as we swing away from the river onto higher ground, the *Gilbertiodendron* gives way to a mixed forest, its canopy gaps delivering light to a clamorous undergrowth of brush, saplings, thorny vines, and woody lianas, through which we climb hunchbacked behind the day's Pygmy point man. The thickest zones of such early-successional vegetation are known in local slang as *kaka zamba*, politely translated as "crappy forest." Today it's Bakembe, younger and stronger than Ndokanda, who cuts us a tunnel through the *kaka*.

The most devilish of the thorny vines is *Haumania dankelmaniana*, mentioned already as a favored gorilla food. Looping high and low throughout the understory, weaving *kaka zamba* into a tropical brier patch, forever finding chances to carve bloody scratches across unprotected ankles and toes, *Haumania* is the bush-whacker's torment. Even a Congo-walker as seasoned as Fay has to spend much of his time looking down, stepping carefully, minimizing the toll on his feet. Of course Fay would be looking down anyway, because that's where so much of the data is found—scat piles, footprints, territorial scrape marks, masticated stems, grouty tracks left by red river hogs nose-plowing through leaf litter, pangolin burrows, aardvark burrows, fallen leaves, fallen fruit. Fay's GPS tells us where we are, while his map and our compasses tell us which way to go. There are no human trails in this forest, because there are no resident humans, few visitors, and no destinations.

Fay pauses over a pile of gorilla shit, recognizing seeds of *Marantes glabra* as a hint about this animal's recent diet. Farther on, he notes the hole where a salt-hungry elephant has dug for minerals. Farther still, the print of a yellow-backed duiker, one of the larger forest antelopes. Each datum goes into the notebook, referenced to the minute of the day, which will be referenced in turn by his GPS to longitude and latitude at three decimal points of precision. Years from now, his intricate database will be capable of placing that very pile of gorilla shit at its exact dot in space-time, should anyone want to know.

When it comes time to ford the Mopo, Fay wades knee-deep into the channel with his video camera pressed to his face. Spotting a dark lump against the white sand, he gropes for it one-handed, still shooting. "Voilà. A palm nut." He shows me the hard, rugose sphere, smaller than a walnut, light in weight but heavy with import. It's probably quite old, he explains. He has found thousands like this in his years of wading the local rivers, and carbon-dating analysis of a sizable sample revealed them to be durable little subfossils, ranging back between 990 and 2,340 years. Presumably they wash into a stream like the Mopo after centuries of shallow burial in the soil nearby. What makes their presence mysterious is that this species of palm, *Elaeis guineensis*, is known mainly as an agricultural species, grown on plantations near traditional Bantu villages at the fringe of the forest and harvested for its oil. *Elaeis guineensis* seems to need cleared land, or at least gaps and edges, and to be incapable of competing in dense, mature forest. The abundance of ancient oil-palm nuts in the river channels suggests a striking possibility: that a vast population of early Bantu agriculturalists once occupied this now vacant and forested region. So goes Fay's line of deduction, anyway. He hypothesizes that those proto-Bantus cut the forest, established palm planta- tions, discarded millions or billions of palm nuts in the process of extracting oil, and then vanished, as mysteriously as the Anasazi vanished from the American Southwest. Some scholars argue that natural climate change over the past three millennia might account for the coming and going of oil palms, the natural ebb and return of forest, but to Fay it doesn't make sense. "What makes sense," he says, "is that people moved in here, grew palm nuts, and then died out." Died out? From *what*? He can only guess: maybe warfare, or a killer drought, or population overshoot leading to eco- logical collapse, or severe social breakdown resulting from some combination of such factors. Or maybe disease. Maybe an early version of AIDS or Ebola or bubonic plague emptied the region of people, more or less abruptly, allowing the forest to regrow. There's no direct evidence for this cataclysmic depopulation, but it's a theme that will recur throughout Fay's hike. Meanwhile, he drops the palm nut into a Ziploc bag.

Just beyond the Mopo, we sneak up on a group of gorillas feed- ing placidly in a *bai*, a boggy clearing amid the forest. We approach within thirty yards of an oblivious female as she works her way through a salad of *Hydrochoris* stems. Fastidiously, she nips off the tender white bases, tossing the rest aside. Her face is long and tranquil, with dark eyes shaded beneath her protrusive brow. The hair on her head is red, Irish red, as it generally is among adult lowland gorillas. Her arms are huge, her hands big and careful. Leaving me behind, Fay skulks closer along the bai's perimeter. When the female raises her head to look straight in his direction, the intensity of her stare seems to bring the whole forest to silence. For a minute or two she looks puzzled, wary, menacingly stern. Then she resumes eating. Fay gets the moment on zoom-lens video. Later he tells me that he froze every muscle while she glow- ered at him, not daring to lower the camera, not daring to move, while a tsetse fly sucked blood from his foot.

The video camera, with its soundtrack for verbal annotations and its date-and-time log, is becoming one of his favorite tools. He shoots footage of major trees, posing a Pygmy among the buttresses for scale. He shoots footage of monitor lizards and big unidentified spiders. He shoots footage, for the hell of it, of me floundering waist-deep in mud. Occasionally he does a slow 360-degree pan to show the wraparound texture of a patch of forest. And when I alert him that a leech has attached itself to one of the sores on his right ankle, he videos that. Then he hands me the camera while Madzou burns the leech off with a lighter, so that I can capture the operation from a better angle.

Just before noon he inspects another fresh mound of elephant dung, poking his finger through the mulchy gobs. Elephants in this forest eat a lot of fallen fruit, but just what's on the menu lately? He picks out seeds of various shape and size, identifying each at a glance, reciting the Latin binomials as he tosses them into a pile: *Panda oliosa, Tridesmos stemon, Antrocaryon klaineana, Duboskia macrocarpum, Tetropleura tetraptera, Drypetes gosweilieri,* and what's this other little thing, can't remember, wait, wait . . . oh yeah, *Treculia africana.* As I squat beside him, impressed by his knowledge and scribbling the names, he adds: "Of course, this is where you get footworms, standing in elephant dung like this."

We make camp along a tributary of the Mopo. The Pygmies erect a roof beam for the main tarp and a log bench for our ease before the campfire. According to the topofil, Madzou reports, our day's progress has been 33,420 feet. Not a long walk, but a full one. After dark, as Fay and Madzou and I sit eating popcorn, there comes a weird, violent, whooshing noise that rises mystifyingly toward crescendo and then crests—as, whoa, an elephant charges through camp, like an invisible freight train with tusks. Sparks explode from the campfire as though someone had dropped in a Roman candle, and the Pygmies dive for safety. Then, as quickly, the elephant is gone. Anybody hurt? No. Dinner is served and the pachyderm in the kitchen is forgotten, just a minor distraction at the end of a typical day on the Megatransect.

Fay spent the late 1980s at a site in southern CAR, gathering data on the resident gorillas. He was particularly curious about their food choices (all gorillas are vegetarian, but their local diets reflect the plant availabilities of a given ecosystem) and their nesting behavior. The lowland gorilla, like the chimpanzee, is known to build sleeping nests from bent or interwoven branches, and with gorillas those nests are sometimes elaborate. Every gorilla above weaning age makes such a nest, simple or fancy, almost every night. By counting nests, therefore, a biologist can estimate gorilla population density; and from nest counts and other evidence left behind as the animals move, inferences can be drawn about group size, demographic composition, and social organization. In other words, a researcher can learn much without even *seeing* gorillas.

One of the methods Fay used was a standard line-transect survey, which involved cutting straight trails through his study area, creating a rectilinear grid, and then walking the trails repeatedly to count and plot nests. Fay's study-site grid, spanning floodplain and lowland forest from the Sangha River to a smaller stream that ran parallel, was just 3.3 miles wide. He could march all through it, gathering data as he went, in a day. Another of his methods, which proved more congenial to his disposition, was what he labeled a "group follow."

He hit upon this technique, from necessity, toward the end of his fieldwork period. The gorillas were skittish. They generally fled from any contact with humans—that is, mutual visibility or intrusive proximity. Earlier on, Fay had spent a lot of effort trying to habituate certain gorilla groups to his presence. That was difficult, he found. But if a group of gorillas was followed and *not* contacted, there was no need for habituation. He could stay near the group indefinitely—out of sight, beyond earshot—and leave them none the wiser while he collected data from their abandoned nests, their dung, and other residual clues. So he started to shadow them that way.

It required keen tracking skills. Fay enlisted those skills in the person of a brilliant Pygmy tracker named Mbutu Clement, a member of the Bambendjellé clan, who became his mentor and friend. With Mbutu's guidance, he would follow a group of gorillas discreetly but persistently for all of one day or several, holding back at enough distance (several hundred yards) to keep them unaware of his presence. Among the clues Mbutu used were chewed-upon stems of *Haumania dankelmaniana*, the thorny creeper, which gorillas find toothsome. Because its tissue oxidizes quickly when exposed to air, a freshly gnawed stem of *Haumania* retains its whitish inner color for only about five minutes; after ten minutes, it has turned black. Fay and Mbutu tried to stay within the five-minute range of a gorilla group without being perceived. Such fastidious tracking allowed Fay to learn what the gorillas had been eating, how many nests they had built, how often they shat, and what their group size, ages, and gender composition might be, while minimizing the chance that he'd spook them.

Near the end of the study, in late 1988, he and Mbutu followed one group for twelve days, dawn to dark each day, resting and eating and walking in synchronic rhythm with the gorillas. From a reading of his eventual dissertation, it seems that "the twelve-day follow," as he called it, was a high point in his academic fieldwork. It was also a foundational bit of experience for what he would later attempt in the Megatransect.

He returned to grad school in St. Louis meaning to write that dissertation, but after a few months he shoved it aside (not to be finished until eight years later) and flew back to Africa, seizing the irresistible distraction of more fieldwork. His new assignment was to do some surveys of forest elephants in northern Congo. He inherited this project from a biologist colleague who had developed the methodology, gotten the grant, and then found himself laid up with a broken back. Fay took over, choosing to focus the survey on three remote, difficult ecosystems: an area near the Gabonese border known as Odzala, a vast swampland to the east known as Likouala aux Herbes, and, farther north, a zone of trackless forest between the Nouabalé and Ndoki Rivers.

Teaming up with an adventuresome Congolese biologist named Marcellin Agnagna, Fay set himself the delectable (to him) task of traversing all three areas on foot. Elephant data would be the purpose and the result, but the bush travel would be its own reward. For the Odzala trek, they began at a town called Mbomo. "People were amazed that we were going to just walk from Mbomo to Tshembe, which in a straight line is like 130 Ks across the forest," Fay recalls. "The villagers thought we were out of our minds." A year later he returned for a second survey trek in the Nouabalé-Ndoki area, where he had found such a wonderland of undisturbed forest that it would eventually, after much determined but deft politicking by Fay and others, become one of Congo's most treasured national parks. By 1994, Fay himself was director of this Nouabalé-Ndoki National Park project, on a management contract between the Republic of Congo government and the Wildlife Conservation Society. He based himself at a village called Bomassa, on the east bank of the upper Sangha River. Although his administrative duties had grown heavy and his political reach had lengthened, he still slid out for a two- or three-week reconnaissance hike whenever the necessity or the excuse arose. And soon after that he began to brainstorm about applying his leg-power approach on a whole different scale.

His widened perspective came literally from the sky: a hundred feet above the canopy in a Cessna 182. Back in St. Louis he had gotten pilot training, and by 1996 he had found grant money to buy the Cessna. He began flying low-altitude excursions over Congo, Gabon, and the neighboring countries, scanning the landscape as though it were a colorful map on his coffee table, taking himself down to the altitude of parrots and hornbills above areas no road had ever crossed. He logged a thousand hours. He saw the real texture of what was out there—the hidden bais where elephants gathered, the thick groves of Marantaceae vegetation representing bounteous gorilla food, the fishing settlements along small rivers, the poachers' camps secreted in the outback, the Bantu villages, and the great zones of forest where neither settlements, camps, nor villages had yet arrived. "Everything came together because of the airplane," Fay says. "It gave me the big picture." The big picture as he soon sketched it was of a single grandiose hike, complemented with overflights for aerial videography, that would seek to embrace, sample, quantify, interconnect, and comprehend as much of the Central African forest as humanly possible. After more than a year of planning, enlisting collaborators (among whom Nick Nichols was crucial, for his great influence at *National Geographic* magazine, for which he was a staff photographer), gathering permissions from governments, selling his vision to sponsors, arranging logistical support, packing, and further flying, he parked the Cessna and started to walk.

On the afternoon of Day 5 we enter Nouabalé-Ndoki National Park, crossing the Ndoki River in dugout canoes, then paddling up a deep blackwater channel through meadows of swaying *Leersia* grass and continuing onward by foot. We spend the last hour before sunset walking through a rainstorm so heavy it fills the trail with a cement-colored flood. On Day 7 we skirt the perimeter of Mbeli Bai, a large clearing much frequented by elephants and gorillas. Fay's first glimpse of this bai back in 1990, he tells me, was an ugly experience: He found six elephant carcasses, some with their tusks already hacked out, others left to rot until the extraction would be easier. The park hadn't yet been decreed, and poaching was rampant. In recent years the situation is much improved. The park has also brought protection to giant trees of the species most valued for timber, such as *Entandrophragma cylindricum*, informally known as *sapelli*, one of the premium African mahoganies. Pointing to a big sapelli, he says, "There's something you wouldn't see on the other side of the river"—that is, west of the park, where selective logging has already combed away the most formidable trees. Later he notes a mighty specimen of *Peracopsis alata*, far more valuable even than sapelli. A log of *Peracopsis* that size is like standing gold, Fay says, worth about $30,000 coming out of the sawmill. Spotting another, he changes his metaphor: "If sapelli is the bread-and-butter around here, *Peracopsis* is the caviar."

We linger through midafternoon with a group of eerily brash chimpanzees, which have gathered at close range to watch us. The chimps hoot and gabble and grunt, perching in trees just overhead, sending down pungent but unmalicious showers of urine, scratching, cooing, thrashing the vines excitedly, ogling us with intense curiosity. One female holds an infant with an amber face and huge, back-lit orange ears, neither mother nor baby showing any fear. A young chimp researcher named Dave Morgan, who has joined us for this leg of the hike, counts eleven individuals, including one with a distinctively notched left earlobe.

It's a mesmerizing encounter, both for us and for them, but after two hours with the chimps we push on, then find ourselves running out of daylight long before we've reached a suitable campsite. None of us wants a night without water. We grope forward in the dark, wearing headlamps now, cutting and twisting through *kaka zamba*, finally stumbling into a sumpy, uneven area beside a muddy trickle, and Fay declares that this will do. Early next morning we hear chimps again, calling near camp. With Morgan's help, we realize that it's probably the group from yesterday, having tracked us and bedded nearby. Camp-following chimps? Aren't they supposed to be terrified of humans, who commonly hunt and eat chimpanzees throughout Central Africa? The sense of weird and unearthly comity only increases when, on Day 8, we cross into an area known as the Goualougo Triangle.

At 4:15 that morning I'm awake in my tent, preparing for the day's walk by duct-taping over the sores and raw spots on my toes, ankles, and heels. To travel the way Mike Fay travels is hard on the feet, even hard on *his* feet, not because of the distance he walks but because of where and how. After a week of crossing swamps and stream channels behind him, I've long since converted to Fay's notion of the optimal trail outfit—river sandals, shorts, one T-shirt that can be rinsed and dried. But the problem of foot care remains, partly because of the unavoidable cuts, stubs, and slashes inflicted by the *Haumania dankelmaniana* vine and other hazards, and partly because the sandy mud of Congolese swamps has an effect like sandpaper socks, chafing the skin away wherever a sandal strap binds against the foot. So I've adopted the practice of painting my feet with iodine every morning and night, and (at the suggestion of another tough Congo trekker, a colleague of Fay's named Steve Blake) using duct tape to cover the old sores and protect against new ones. The stuff holds amazingly well through a day of swamp-slogging, and although peeling off the first batch isn't fun, removal becomes easier on later evenings when there's no more hair on your feet. Since I've got a small roll of supple green tape as well as a larger roll of the traditional (but stiffer, less comfortable) silver, I even find myself patterning the colors—green crosses over the tops of the feet, green on the heels, silver on the toes: a fashion statement. If my supplies of iodine and tape can be stretched for another ten days and my mental balance doesn't tip much further, I'll be fine.

At 4:30 A.M. I hear Dave Morgan, awake now in the tent beside mine, beginning to duct-tape his feet.

Over breakfast, Fay himself asks to borrow my tape for a few patches on his toes and heels. I give him the silver, selfishly hoarding the green. Then again we walk.

Demarcated by the Goualougo River on one side, the Ndoki River on another, the Goualougo Triangle is a wedge-shaped area extending southward from the southern boundary of Nouabalé-Ndoki National Park. In other words, it's ecologically continuous with the park but not part of it statutorily, and isolated from the wider world by the two rivers. Having already made our Ndoki crossing, we enter on solid ground from the park.

The Triangle embraces roughly 300 square kilometers of primary forest, including much excellent chimpanzee habitat, a warren of elephant trails, and an untold number of big sapelli trees, all encompassed within a logging concession held by a company called Congolaise Industrielle des Bois (CIB), the largest surviving timber enterprise in northeastern Congo. With two sawmills, a shipyard, a community hospital, and logging crews in the forest, CIB employs about 1,200 people, mostly in the towns of Kabo and Pokola, along the Sangha River. Although the company has shown willingness to collaborate with WCS on management of a peripheral zone south of the park, especially toward restricting the commercial trade in bushmeat (wild species killed for food) coming out of the forest, tension now seems to be gathering around the issue of the Goualougo Triangle. Mike Fay originally hoped to see that wedge of precious landscape included in the park, but when the boundaries were drawn, in 1993, the Goualougo was lined out. About the same time, CIB acquired the concession from another logging company that went into receivership. After a half-decade of benign inattention, CIB now wants to move toward logging the Goualougo, or at least to conduct an on-the-ground assessment of the timber resource and the costs of extracting it. That assessment—a *prospection*, in the jargon of Francophone forestry—

will put a price tag on the Triangle. Meanwhile the company, in a spirit that mixes cooperation with hardheaded bargaining, has invited WCS to do a parallel prospection, theirs to assess the area's biological value. Weeks after returning from the Congo, I hear CIB's position on the Goualougo put by the company's president, Dr. Hinrich Stoll. "You cannot just say, 'Forget about it, it is completely protected,'" he tells me by phone from his office in Bremen, Germany. "We all want to know how much it is worth." Once its worth has been gauged, both in economic and in biological terms, also in social ones, then perhaps the international community of conservationists and donors will see fit to compensate his company—yes, and the working people of Pokola and Kabo, Dr. Stoll stresses—for what they're being asked to give up.

But that talk of compensation, of balancing value against value, of ransoming some of the world's last ingenuous chimpanzees, comes later. As I stroll through the Goualougo with Fay, he turns the day into a walking seminar in forest botany, instructing me or quizzing Madzou and Morgan on the identity of this tree or that. Here's an *Entandrophragma utile*, slightly more valuable but far less common than its congeneric *Entandrophragma cylindricum*. Its fruits resemble blackish yams festooned with wiry little roots, not to be confused with the banana-shaped fruit of another *Entandrophragma* species, *candoliae*. And here's still another, *Entandrophragma angolense*. What about that tree there—what is it, Morgan? he demands. Um, an *Entandrophragma*? Wrong, Fay says, that one's *Gambaya lacourtiana*. Of course to me these are all just huge hulking boles, thirty feet around, rising to crowns in the canopy so high that I can't even see the shapes of their leaves. Morgan and Madzou are earnest students. Fay is a stern but effective teacher, sardonic one moment, lucid and helpful the next, drawing tirelessly on his own encyclopedic knowledge and his love for the living architecture of the forest. Now he directs Morgan's attention to the fine, fissured, unflaky bark of *Gambaya lacourtiana*, which is not to be confused with the more subtly fissured bark of *Combretodendron macrocarpum*, which is not to be confused with . . . a pile of lumber awaiting shipment from Kabo.

The good news from Day 8 is that Fay finds no *Peracopsis alata*, no standing gold, no caviar, at least along this line of march in the Goualougo Triangle. The bad news is that there's an abundance of *Entandrophragma*, CIB's bread-and-butter. By the time the prospection team arrives to confirm or modify those impressions, Fay will be somewhere else, continuing his own singular sort of prospection at his own pace and scale.

From the Goualougo Triangle we make our way upstream along the Goualougo River, crossing back into the park. On the evening of Day 11 we're settled near an idyllic little bathing hole, a knee-deep pool with a sand bottom and a fallen log nearby that makes a good shelf for my bottle of Dr. Bronner's soap. Peeling away my duct-tape socks after a gentle soak underwater, I feel exquisite relief. I wash my feet carefully, the rest of my body quickly, and then, given the luxury of deep clear water, my hair. I rinse my shorts and T-shirt, wring them, put them back on. It's been a good day, enlivened by another two-hour encounter with a group of fearless chimps. For dinner there'll be a pasty concoction known as *foufou*, made from manioc flour and topped with some kind of sauce, plus maybe a handful of dried apricots for dessert. Then a night's blissful sleep on the ground; then fresh duct tape; then another day's walk. Having fallen into his rhythm, I've begun to see why Mike Fay loves this perverse, unrelenting forest so dearly.

Seated beside the campfire, Fay puts Neosporin antiseptic on his ragged toes. Several footworms have burrowed in there and died, mortally disappointed that he wasn't an elephant. The ointment, as he smears it around, mixes with stray splatters of mud to make an unguent gray glaze. No, he affirms, there's no escaping foot hassles out here. You've just got to keep up the maintenance and try to avoid infection. When necessary, you stop walking for a few days. Lay up, rest. Let them heal. Wait it out.

So he says. I can scarcely imagine what Fay's feet might have to look like before he resigns himself to that.

At the end of Day 13 we make camp on a thickly forested bench above the headwaters of the Goualougo, which up here is just a step-across stream. Our distance traversed since morning, as measured by the Fieldranger, is 42,691 feet. Our position is 2E26.297′

north by 16E36.809′ east, which means little to me but much to the great continuum of data. This particular day, alas and hoorah, has been my final one of walking with Fay, at least for now. (The plan is that I'll return months later to share other legs of the hike.) Tomorrow I'll point myself toward civilization, retracing our trail of string and machete cuts to the Sangha River. Morgan and three of the Pygmies will accompany me.

And Fay? He'll continue northeastward to the rendezvous with Nick, then loop down again through Nouabalé-Ndoki National Park before heading out across the CIB logging concessions and the other variously tracked and untracked forests of Central Africa. The Megatransect has only begun: thirteen days gone, roughly four hundred to go. Many field notes remain to be taken, many video- and audiotapes to be filled, much data to be entered in the computers, many kilometers of topofil to unroll. Then will come the challenge of making it all matter—collation, analysis, politics. When he reaches the seacoast of Gabon, Fay has told me, he'll probably wish he could just turn around and start walking back.

II. THE GREEN ABYSS
March–July 2000

It takes a hardheaded person to walk 2,000 miles across west-central Africa, transecting all the wildest forests remaining between a northeastern corner of the Republic of Congo and the Atlantic. It takes a harder head still to conceive of covering that terrain in a single, sustained, expeditionary trudge. There are rivers to be ferried or bridged, swamps to be waded, ravines to be crossed, vast thickets to be carved through by machete, and one tense national border, as well as some lesser impediments—thorny vines, biting flies, stinging ants, ticks, vipers, tent-eating termites, and the occasional armed poacher. As though that weren't enough, there's a beautifully spooky forest about midway on the route that's believed to harbor Ebola virus, the cause of lethal epidemics in

nearby villages within recent years. The logistical costs of an enterprise on this scale, counting high-tech data-gathering gizmos and
aerial support, can run to hundreds of thousands of dollars. The
human costs include fatigue, hunger, loneliness, tedium, some
diseases less mysterious than Ebola, and the inescapable nuisance
of infected feet. It takes an obdurate self-confidence to begin such
a journey, let alone finish. It takes an unquenchable curiosity and
a monomaniacal sense of purpose.

J. Michael Fay is as obdurate and purposeful as they come. But
even for him there arrived a moment, after eight months of walking, when it looked as if the whole venture would end sadly. One of
his forest crew, a young Bambendjellé Pygmy named Mouko, lay
fevering on the verge of death. Hepatitis was taking him down fast.

Mouko's illness was only the latest travail. Within recent days
Fay had been forced to backtrack around an impassable swamp.
His twelve Bambendjellé crewmen, even the healthy ones, were
exhausted and ready to quit. That border crossing, which loomed
just ahead, had begun to appear politically problematic—no Gabonese visas to be had for a gang of Congolese Pygmies. And then a
certain Muslim trader went missing between villages along one of
the few human footpaths with which Fay's route converged; as
authorities reacted to the man's disappearance, Fay began dread-
ing the prospect that he and his feral band might come under suspicion and be sidetracked for interrogation. Suspending the march
to nurse Mouko, he found himself stuck in a village with bad water.
He was running short of food, with not even enough pocket money
to buy local bananas. The Megatransect was in megatrouble.

If Mouko dies, Fay thought, it's probably time to roll up the tents
and capitulate. He would abandon his dream of amassing a great
multidimensional filament of forest-survey data, continuous both
in space and in time. He would stop recording all those little particulars—the relative freshness of every pile of elephant dung, the
location of every chimp nest and aardvark burrow, the species and
girth of every big tree—in the latest of his many yellow notebooks.
He would stop walking. Human exigencies would preempt meth-
odological imperatives and vaulting aspirations. If Mouko dies, he
figured, I'll drop everything and take the body home.

Even from the start, in late September of last year, it looked like a daunting endeavor, far too arduous and demented to tempt an ordinary tropical biologist, let alone a normal human being. But Fay isn't ordinary. By his standards, the first three months of walking were a lark. Then the going got sticky.

Having crossed Nouabalé-Ndoki National Park and that stunning wedge of pristine forest known as the Goualougo Triangle, having hiked south through the trail-gridded timber concessions and boomtown logging camps of the lower Ndoki watershed, Fay and his team angled west, toward a zone of wilderness between the Sangha and the Lengoué Rivers, both of which drain south to the main stem of the Congo River. What was out there? No villages, no roads. On the national map it was just a smear of green. Fay traveled along elephant trails when possible, and when there were none, he bushwhacked, directing his point man to cut a compass-line path by machete.

A strong-armed and equable Pygmy named Mambeleme had laid permanent claim to the point-man job. Behind him walked Fay with his yellow notebook and video camera, followed closely by Yves Constant Madzou, the young Congolese biologist serving as his scientific apprentice. Farther back, beyond earshot so as not to spook animals, came the noisier and more heavily burdened entourage—twelve Pygmy porters and a Bakwele Bantu named Jean Gouomoth, nicknamed Fafa, Fay's all-purpose expedition sergeant and camp cook. They had proceeded that way for many weeks, in a good rhythm, making reasonable distance for reasonable exertion, when gradually they found themselves submerged in a swale of vegetation unlike anything Fay had ever seen.

Trained as a botanist long before he did his doctoral dissertation on gorillas, Fay describes it as "a solid sea of Marantaceae"—the family Marantaceae constituting a group of herbaceous tropical plants that includes gangly species such as *Haumania dankelmaniana*, the thorny ankle-ripping nuisance, and its near cousin *Haumania liebrechtsiana*, a more vertically inclined plant that can grow into stultifying thickets, denser than sugarcane, denser than grass, dense as the fur on a duck dog. The Marantaceae brake that Fay and his team had now entered, just east of the Sangha

River, stretched westward for God only knew how far. Fay himself, with a GPS unit and a half-decent map but no godlike perspective, knew not. All he could do was point Mambeleme into the stuff, like a human Weedwacker, and fall in behind.

Sometimes they moved only sixty steps an hour. During one ten-hour day they made less than a mile. The green stems stood fifteen feet high, with multiple branches groping crosswise and upward, big leaves turned greedily toward the sun. "It's an environment which is completely claustrophobic," Fay says later, from the comfort of retrospect. "It's like digging a tunnel, except there is sunlight." The cut stems scratched at their bare arms and legs. Sizable trees, offering shade, harboring monkeys, were few. Flowing water was rare, and each afternoon they searched urgently for some drinkable sump beside which to camp. When they did stop, it took an hour of further cutting just to clear space for the tents.

On the march, Fay spent much of his time bent at the waist, crouching through Mambeleme's tunnel. He learned to summon a Zenlike state of self-control, patience, humility. The alternative was to start hating every stem of this Marantaceae hell, regretting he had ever blundered into it—and along that route a person might go completely nuts. Mambeleme and the other Pygmies had their own form of Zenlike accommodation. "*Eyali djama,*" they would say. "*Njamba eyaliboyé.*" That's the forest. That's the way it is.

But this wasn't the real forest, woody and canopied and diverse, that Mike Fay had set out to explore. It was something else, an awesome expanse of reedy sameness. Later he named it the Green Abyss.

They reached the Sangha River, crossed in borrowed pirogues, then plunged westward into more of the same stuff. Fay had flown this whole route in his Cessna, scouting it carefully, but even at low elevation he hadn't grasped the difficulty of getting through on foot. Villagers on the Sangha, whose own hunting and fishing explorations had taught them to steer clear of that trackless mess, warned him: "It's impossible. You cannot do it. You will fail. You will be back here soon." Fay's response was: "We have maps. We have a compass, and we have strong white-man medicine. We will make it." He was right. But it took ten miserable weeks. Having spent

New Year's Eve in the Green Abyss, he wouldn't emerge until early March.

"We drank swamp water for three weeks in a row. We did not see any flowing water for almost a month," Fay recalls. "Miraculously, we only had one night where we had to drink water out of a mud-hole." It was an old termite mound, excavated by an aardvark or some other insectivore and lately filled with rainwater. The water was thick with suspended clay, grayish brown like latte but tasting more like milk of magnesia.

Food was another problem, since their most recent rendezvous with Fay's logistical support man, an ever-reliable Japanese ecologist named Tomo Nishihara, had been back at the Sangha; they were now days behind schedule and would be on starveling rations long before they reached the next resupply point. So by satellite phone Fay and Tomo arranged an airdrop: twenty-kilogram bags of manioc and fifty-can cases of sardines dumped without parachutes from a low-flying plane. The drop was a success, despite one parcel's ripping open on a tree limb, leaving a plume of powdered manioc to sift down like snow and fifty sardine cans mooshed together like a crashed Corvair. The men binged on the open sardines, then resumed walking.

Other problems were less easily solved. There were tensions and deep glooms. There were days that passed into weeks not just without flowing water but without civil conversation. Not everyone on the team found his own variant of *Njamba eyaliboyé*. By the time they reached the Lengoué River, Yves Madzou had had enough, and Fay had had enough of his enoughness. By mutual agreement, Yves left the Megatransect to pursue, as the saying goes, other interests. He was human, after all.

Fay was Fay. He marched on.

After six months, Fay and his crew paused for rest and resupply at a field camp called Ekania, on the upper Mambili River, within another spectacular area of Congolese landscape, Odzala National Park. Odzala is noted for its big populations of forest elephants and gorillas, which show themselves in the small meadowy clearings known as bais, sparsely polka-dotting the forest. Mineral salts, edible sedges, and other toothsome vegetation at a bai attract not just elephants and gorillas but also forest buffalo, sitatungas, bongos, and red river hogs, sometimes in large groups. Of course Fay wanted to visit the bais, which he had scouted by plane but never explored on foot; he also wanted to take the measure of the forest around them.

Odzala's elephants suffered heavily from poaching during the late 1980s and early '90s, until a conservation program known as ECOFAC, funded by the European Commission, assumed responsibility for managing the park, with a stringent campaign of guard patrols and a guard post on the lower Mambili to choke off the ivory traffic coming downriver. Access deep into Odzala along the Mambili, a chocolaty stream whose upper reaches are narrow and strained by many fallen trees, is still allowed for innocent travelers not carrying tusks. That's how Tomo brought the resupply crates up to Ekania. It was a ten-hour trip by motorized dugout from the nearest grass airstrip, and on this occasion I traveled with him.

Fay, bare-chested and walnut brown, with a wilder mane of graying hair than I remembered, stood on a thatched veranda taking video of us as we docked. Without pulling the camera from his eye, he waved. I can't remember if I waved back; more likely I saluted. He had begun to remind me of a half-mad, half-brilliant military commander gone AWOL into wars of his own choosing, with an army of tattered acolytes attending him slavishly—rather like Brando's version of Conrad's Kurtz in *Apocalypse Now*, only much skinnier.

It was the first time I'd seen Fay since Day 13 of the Megatransect, back in October, when I split off from his forest trek and walked out to a road. Now his shoulder bones stood up like the knobbled back of a wooden chair, suggesting that he'd lost twenty or thirty pounds. But his legs were the legs of a marathoner. The quiet, clinical smile still lurked behind his wire glasses. Greeting him again here on Day 182, many hundreds of miles deep in the equatorial outback, I felt like Stanley addressing Dr. Livingstone.

"Every day that I walk," Fay volunteered, "I'm just happier that I did the Megatransect." He said "did" rather than "am doing," I noticed, though in fact he was only halfway along. Why? Because

the advance planning and selling phase had been the most oner-
ous part, I suspected, after which the actual walk felt like raking
in a poker-game pot. Aside from a chest cold and a few footworm
infections, and notwithstanding the weight loss, he had stayed
healthy. His body seemed to have reached some sort of equilibrium
with the rigors of the forest, he said; his feet, I saw, were marked
with pinkish scar tissue and pale sandal-strap bands against the
weathered brown. No malaria flare-ups, no yellow fever. Just as
important, he was having fun—most of the time, anyway. He
described his ten weeks in the Green Abyss, making clear that *that*
passage, far from fun, had been "the most trying thing I've ever
done in my life." But now he was in Odzala, lovely Odzala, where
the bongo and the buffalo roam. He had a new field companion to
help with the botany, a jovial Congolese man named Gregoire
Kossa-Kossa, forest-hardy and consummately knowledgeable, on
loan from the Ministry of Forestry and Fishing. Fafa, his crew boss
and cook, had grown into a larger role, which included data-
gathering chores earlier handled by Yves. And his point man,
Mambeleme, now with a buffed-out right arm and a machete so
often sharpened that it was almost used up, had proven himself a
champion among trail-cutters. The rest of Fay's crew, including
the brothers Kati and Mouko, had suffered badly from that chest
cold they all caught during a village stop but now seemed fine.
Mouko's more serious illness, along with other tribulations, was
yet to come.

Meanwhile Fay's own data gathering had continued, providing
some new and significant impressions of Odzala National Park. For
instance, one day in a remote floodplain forest, Fay, along with
Mambeleme and Kossa-Kossa, had sighted a black colobus monkey,
the first report of that rare species within the park. In the famed
bais of Odzala he saw plenty of elephants, as he had expected, but
during his long cross-country traverses between one bai and
another he found a notable absence of elephant trails and dung,
suggesting that a person shouldn't extrapolate from those bais to
an assumption of overall elephant abundance. His elephant-sign
tallies, recorded methodically in the current yellow notebook,
would complement observations of elephant distribution made by
ECOFAC researchers.

Maybe those notebooks would yield other insights too. Maybe the Megatransect wasn't just an athletic publicity stunt, as his critics had claimed. It occurred to me as an intriguing possibility, not for the first time, that maybe Mike Fay wasn't as crazy as he looked.

After a few days at Ekania we set off toward the Mambili headwaters and a large bai called Maya North, near which was another ECOFAC field camp used by elephant researchers and visiting film crews. The usual route to Maya North camp was upriver along the Mambili, traveling some hours by motorized dugout to a point where ECOFAC workers had cut a good trail. We came the back way, bushwhacking on an overland diagonal. That evening, as we sat by the campfire trading chitchat with several Congolese camp workers, the talk turned to boat travel on the upper Mambili. Well, we didn't use a boat, Fay mentioned. *You didn't?* they wondered. *Then how did you get here?* We walked, Fay said. *Walked? All the way from Ekania? There's no trail.* True but irrelevant, Fay said.

At daybreak on Day 188 we were at the bai, watching eighteen elephants in the fresh light of dawn as they drank and groped for minerals in the stream. Some distance from the others stood an ancient female, emaciated, failing, her skull and pelvic bones draped starkly with slack gray skin. Amid the herd was a massive bull, who swept his raised trunk back and forth like a periscope, tasting the air vigilantly for unwelcome scents. He caught ours. There was a subtle shift in mood, then the bull initiated a deliberate, wary leave-taking. One elephant after another waded off toward the far side of the bai, disappearing into the trees. By sunup they were gone.

By midday so we were, walking on.

From the upper Mambili, Fay planned to ascend toward an escarpment that forms the divide between the Congo River basin and a lesser system, the Ogooué, which drains to the Atlantic through Gabon. I would peel off again on Day 195, using another resupply rendezvous with Tomo as my chance to exit. As it happened, Tomo needed three boatmen and a chainsaw to get his load of supplies that far up the snag-choked Mambili, but going back downriver would be easier, and we figured to reach the airstrip in two days.

On the morning of the day of my departure, Fafa was laid flat by a malarial fever, so Fay himself oversaw the sorting and packing of new supplies: sacks of manioc and rice and sugar, cans of peanut butter and sardines, bundles of salted fish, big plastic canisters of pepper and dried onions, cooking oil, granola bars, freeze-dried meats, cigarettes for the crew, many double-A batteries, a fresh stack of colorful plastic bowls, and one package of seaweed, recommended by Tomo as a complement to the salted fish. Finally the packs were ready, the tents struck; Fafa rallied from his fever, and I walked along behind Fay and Mambeleme into the early afternoon.

Fay and I had agreed where I would rejoin him next: at an extraordinary set of granite domes, known as *inselbergs* ("island mountains"), that rise up like huge stony gumdrops from a forest in northeastern Gabon. The forest, called Minkébé, is ecologically rich but microbially menacing; many months earlier, as we had knelt over my map on the floor of an office at the National Geographic Society in Washington, this was where Fay had written "Ebola region" in red ink. "We'll meet you on the other side of the continental divide," he told me cheerily now. "On our way to the Atlantic Ocean."

Backtracking on the trail to catch Tomo's boat, I shook hands with Kossa-Kossa, Fafa, and each of the Pygmy crew, thanking them for their good company and support. I was fascinated by these rough-and-ready Bambendjellés, whom Fay had somehow cajoled and bullied across hundreds of miles, leading them so far from their home forest into an alien landscape, an alien realm of experiences. They had been challenged beyond imagining, stressed fearfully, but so far they hadn't broken; they put me in mind of the sort of Portuguese seamen, uneducated, trusting, adaptable, who must have sailed with Ferdinand Magellan. By way of farewell, I told them in bad Lingala: *"Na kotala yo, na sanza mibalé."* I'll see you in two months.

I was wrong. It would be three months before Fay reached the inselbergs, an interval encompassing some of his most hellish times since the Green Abyss. And when I did rejoin him there, Mambeleme and all the others would be gone.

Fay and his team followed the escarpment northward along its crest, a great uplifted rim that may have once marked the bank of an ancient body of water. Kossa-Kossa left the troop, as planned, to return to his real-life duties. The others shifted direction again, heading into a thumb of territory where the Republic of Congo obtrudes westward against Gabon. They struck toward the Ouaga River and found it defended by a huge swamp, which at first seemed passable but grew uglier as they committed themselves deeper. By insidious degrees, it became a nightmare of raffia palms and giant pandanus standing in four feet of black water and mud, the long pandanus leaves armed with rows of what Fay recalls as "horrid cat-claw spines." He and the crew spent two nights there in a small cluster of trees, among which they built elevated log platforms to hold their tents above the muck. Pushing forward, Fay saw the route get worse: deeper water, no trees, only more raffia and cat-claw pandanus, and five days' distance of such slogging still ahead, with a chance that any rainstorm would raise the water and trap them. Finally he ordered retreat, a rare thing for Fay, and resigned himself to a long detour through a zone for which he had no map.

After circumventing the Ouaga swamp, they converged with a human trail, a simple forest footpath that serves as an important highway linking villages in that northwestern Congo thumb. The footpath took them to a village called Poumba, where they picked up two pieces of bad news: that the Gabonese border crossing would be difficult at best, due to festering discord between local authorities on the two sides, and that a Muslim trader who dealt in gold and ivory had vanished along the footpath under circumstances suggesting foul play. From a certain perspective (one that the local gendarmerie might well embrace), the trader's disappearance coincided suspiciously with another bit of odd news: that a white man with an entourage of Pygmies had materialized from the forest on a transcontinental stroll to count aardvark burrows and elephant dung (so he claimed) and was making fast tracks for the Gabonese border. It could look very incriminating, Fay knew. He felt both eager to move and reluctant to seem panicky. Added to those concerns was another, seemingly minor. For the third time in two weeks one of the Pygmies, Mouko this time, seemed to be suffering from malaria. But a dose of Quinimax would fix that, Fay thought.

Over the next few days Mouko got weaker. He couldn't lug his pack. At times he couldn't even walk and had to be carried. Evidently it was hepatitis, not malaria, since his urine was dark, the Quinimax brought no improvement, and his eyes were going yellow. Fay slowed the pace and took a turn carrying Mouko's pack. Hiding his uncertainty, he wondered what to do. *All the Pygmies think Mouko is going to die now*, he wrote in his notebook on Day 241. Mouko seemed languid as well as sick, with little will to live, while the others had already turned fatalistic about his death. Fay became Mouko's chief nurse. He scolded the crew for sharing Mouko's manioc, using his plate, making cuts on his back to bleed him, and various other careless or well-meant practices that could spread the infection. To the notebook, Fay confided: *I am so sick and tired of being the parent of 13 children, it is too much. Thank god I never had children—way too much of a burden. Solo is the way to go—depend on yourself only. The trouble in a group like this is it's like you're an organism. If one part of you is sick or lost the whole organism suffers.* For another ten days after that entry, Mouko's survival remained in doubt.

They pushed toward Garabinzam, a village near the west end of the footpath, on a navigable tributary of the Ivindo River, which drains into Gabon. On the last day of walking to Garabinzam, the team covered nine miles, Kati carrying his brother Mouko piggyback for most of the way. That evening, Fay wrote: *I need to ship these boys home. You can just tell they are haggard, totally worn out. No matter how good they were they are just going to go down one by one. I would love to keep my friends but I would be betraying them if I made them stay on any longer—it would be unjust.*

Several days later, he departed from his line of march—and from all his resolutions about continuity—to evacuate Mouko downriver by boat. They would try for a village at the Ivindo confluence, on the Gabonese side; from there, if Mouko survived, he could be moved to a hospital in the town of Makokou. Fafa would meanwhile escort the others back to their home forest, hundreds of miles east, sparing them from the onward trudge and the unwelcoming border. Fay himself would pick up the hike in Gabon. One stretch of the planned route would remain unwalked—roughly eighteen miles, from Garabinzam overland to the border—as a rankling gap in the data set, a blemish on the grand enterprise, and a token (this is my view, not his) of Fay's humanity.

Left Garabinzam, all is well, he wrote briskly on May 24, 2000, which in Megatransect numeration was Day 248. But he also wrote, almost plaintively: *Pygmies didn't say goodbye.*

Mouko survived and went home. Starting from scratch, Fay gathered a new crew from the villages and gold-mining camps of the upper Ivindo region. He found an able young Pygmy named Bebe, with good ears for wildlife and a strong machete arm, who emerged before long as his new point man; he found a new cook and eight other forest-tough Pygmy and Bantu men; he found energy, even enthusiasm, to continue. They set off on a long arc through the Minkébé forest, targeting various points of interest, most dramatic of which were the inselbergs. That's where I next see Fay, on Day 292, when Tomo and I step out of a chartered helicopter that has landed precariously on one of the smaller mounds.

Skin browner, hair longer and whiter, Fay looks otherwise unchanged. Same pair of river shorts, same sandals, same dry little smile. I have brought him three pounds of freshly ground coffee and a copy of Michael Herr's *Dispatches,* another of the Vietnam war memoirs that he finds fascinating. If he's pleased to see me, for the company, for the coffee, he gives no sign.

At once he begins talking about data. He has been finding some interesting trends. For instance, regarding the gorillas. It's true, he says, picking up a discussion from months earlier, that there's a notable absence of gorillas in the Minkébé forest. Since crossing the border, he hasn't heard a single chest-beat display and he has seen only one pile of gorilla dung. Back in Odzala National Park, over a similar stretch, he would have counted three or four *hundred* dung piles.* Elephants are abundant; duikers and monkeys and pigs, abundant. But the gorillas are missing. He suspects they were wiped out by Ebola.

---

* This was true when Fay said it; no longer. During 2002 and 2003, Ebola struck also in Odzala and roundabout, killing dozens of people in villages and hundreds—probably thousands—of gorillas in the forest. But that's part of another story for another time.

The Minkébé forest block, encompassing more than 12,500 square miles of northeastern Gabon, represents one of the great zones of wilderness remaining in Central Africa. Much of it stands threatened by logging operations, bushmeat extraction such as inevitably accompanies logging, and elephant poaching for ivory. But the Gabonese government has recently taken the admirable step of designating a sizable fraction (2,169 square miles) of that block as the Minkébé Reserve, a protected area; and now, in addition, three large adjacent parcels are being considered for possible inclusion.* The Gabonese Ministry of Water and Forests, with technical help and gentle coaxing from the World Wildlife Fund, has been studying the farsighted idea that an enlarged Minkébé Reserve might be valuable not just in ecological terms but also in economic ones for its role in the sequestration of carbon. With greenhouse gases and climate change becoming ever more conspicuous as a global concern, maybe other nations and interested parties might soon be willing to compensate Gabon—so goes the logic—for maintaining vast uncombusted carbon storehouses such as Minkébé.

But before the reserve extension can be approved, on-the-ground assessments must be made. So in the past several years a small group of scientists and forest workers made reconnaissance expeditions into Minkébé—both the original reserve and the proposed extension. They found spectacular zones of forest and swamp, stunning inselbergs, networks of streams, all rich with species and virtually untouched by human presence. They also found—as Mike Fay has been finding—a nearly total absence of gorillas and chimpanzees.

It wasn't always so. In 1984 a paper appeared in the *American Journal of Primatology*, by Caroline Tutin and Michel Fernandez, in which the authors described their census of gorilla and chimpanzee populations throughout Gabon. Using a combination of field transects, habitat analysis, and cautious extrapolation, Tutin and Fernandez estimated that at least 4,171 gorillas lived within

---

* Still more recently, it has become Minkébé National Park, one of thirteen new national parks decreed by President El Hadj Omar Bongo in 2002, partly as a result of Fay's Megatransect.

the Minkébé sector, representing a modest but significant population density. Something seems to have happened to those apes between 1984 and now.

It may have happened abruptly in the mid-1990s, when three Ebola epidemics burned through villages and gold camps at the Minkébé periphery, killing dozens of humans. One of those outbreaks occurred in early 1996 at a village called Mayibout 2, on the upper Ivindo River. It began with a chimpanzee carcass, found dead in the forest and brought to the village as food. Eighteen people who helped with the skinning, the butchering, and the handling of the chimp flesh became sick. Suffering variously from fever, headache, and bloody diarrhea, they were evacuated downriver to the Makokou hospital. Four of them died quickly. A fifth escaped from the hospital, went back to Mayibout 2, and died there. That victim was buried in the traditional way—ceremonies were performed, and no special precautions were taken against infection.

This bare record of facts and numbers comes from a report published three years later, by Dr. Alain-Jean Georges and a long list of coauthors, in a special supplement to *The Journal of Infectious Diseases*. Although the raw chimp flesh had been infectious, the cooked meat evidently hadn't been; no one got sick, the Georges paper asserted, simply from eating it. But once the outbreak began, there were some secondary cases, one human victim infecting another, and the disease spread from Mayibout 2 to a couple villages nearby, Mayibout 1 and Mvadi. By early March, thirty-one people had fallen ill, of whom twenty-one died, for a mortality rate of almost 68 percent. Then it was over, as abruptly as it started. Around the same time, according to later accounts, dead gorillas were seen in the forest.

Mike Fay isn't the only person inclined to connect Minkébé's shortage of gorillas with Ebola virus. Down in the Gabonese capital, Libreville, I heard the same idea from a lanky Dutchman named Bas Huijbregts, associated with the World Wildlife Fund's Minkébé Project, who made some of those reconnaissance hikes through the Minkébé forest, gathering both quantified field data and anecdotal testimony. Gorilla nests, Huijbregts reported, were drastically less abundant than they had been a decade earlier. About the gorillas themselves, he said: "If you talk to all the fishermen, hunters, gold miners, they all have a similar story. Before there were many—and then they started dying off." The apparent population collapse, not just of gorillas but of chimps too, seemed to coincide with the human epidemics. In a hunting camp just north of the Gabonese border, someone showed Huijbregts the grave of a man who, so it was said, had died after eating flesh from a gorilla he had found dead in the forest.

I spoke also with Sally Lahm, an American ecologist who has worked in the region for almost twenty years. Lahm has focused especially on the mining camps of the upper Ivindo, where gold comes as precious flecks from buried stream sediments and protein comes as bushmeat from the forest. Her studies of wildlife and its uses by humans, plus the epidemic events of the mid-1990s, have led her toward the subject of Ebola. When the third outbreak occurred, at a logging camp southwest of Minkébé, she went there with several medical people from the Makokou hospital and played a double role, as both nurse and researcher.

"I'm scared to death of Ebola, because I've seen what it can do," Lahm told me. "I've seen it kill people—up close." Fearful or not, she is engrossed by the scientific questions. Where does Ebola lurk between outbreaks? What species in the forest—a small mammal? an insect?—serves as its reservoir host? How does its ecology intersect the ecology of hunters, villagers, miners? So far, nobody knows.

"It's not a purely human disease," Lahm said. "Humans are the last in the chain of events. I think we should be looking at it as a wildlife-human disease." Besides doing systematic field research, she has gathered testimony from hunters, gold miners, survivors of Mayibout 2. She has also made field collections of tissue from a whole range of reservoir-candidate species, shipping her specimens off to a virology institute in South Africa for analysis. And she has grown suspicious of one particular species that may be the main transfer agent between the reservoir host and humans, but she declined to tell me what species that is. She needs to do further work, she explained, before further talk.

On the evening of Day 299, at Fay's campfire, I hear more on this subject from one of his crewmen, an affable French-speaking Bantu named Thony M'both. *Mayibout deux?* Yes, he was there; he recalls the epidemic well. Yes, it began with the chimpanzee. Some boys had gone hunting with their dogs; they were after porcupine, and they found the chimp, already dead. No, they didn't claim they had killed it. The body was rotten, belly swelling, anyone could tell. Many people helped butcher and cook it. Cook it how? In a normal African sauce. All who ate the meat or touched it got sick, according to Thony. Vomiting and diarrhea. Eleven victims were taken downriver to the hospital—only that many, since there wasn't enough fuel to carry everyone. Eighteen stayed in the village, died there, were buried there. Doctors came up from Franceville (in southern Gabon, site of a medical research institute) wearing their white suits and helmets, but so far as Thony could see, they didn't save anyone. His friend Sophiano Etouck lost six family members, including his sister-in-law and three nieces. Sophiano (another of Fay's crew, also here at the campfire) held one niece in his arms as she died, yet he didn't get sick. Nor did Thony himself. He hadn't partaken of the chimp stew. He doesn't eat chimpanzee or gorilla, Thony avers, implying that's by culinary scruple, not from fear of infection. Nowadays in Mayibout 2, however, *nobody* eats chimpanzee. All the boys who went porcupine hunting that day, they all died, yes. The dogs? No, the dogs didn't die.

The campfire chatter around us has stilled. Sophiano himself, a severe-looking Bantu gold miner with a bodybuilder's physique, a black goatee, a sweet disposition, and an anguished stutter, sits quietly while Thony tells the tale.

I ask one final question: Had you ever before seen such a disease? I'm remembering what I've read about the horrible, chain-reaction Ebola episodes, with victims bleeding profusely, organ shutdown, chaotic hospital conditions, and desperate efforts to nurse or mop up, leading only to further infection. "No," Thony answers blandly. "This was the first time."

Thony's body count differs from the careful report in *The Journal of Infectious Diseases*, so do some other particulars, yet his eyewitness testimony seems utterly real. He's as scared of Ebola as anybody. If he were inventing, he wouldn't invent the chimpanzee's swollen belly. Added to it all, though, is one fact or factoid that he let drop on the first evening I met him—a detail so garish, so perfectly dramatic, that even having heard it from his lips, I'm unsure whether to take it literally. Around the same time as the Mayibout epidemic, Thony told me, he and Sophiano saw a whole pile of gorillas, thirteen of them, lying dead in the forest.

Anecdotal testimony, even from eyewitnesses, tends to be shimmery, inexact, unreliable. To say *thirteen dead gorillas* might actually mean a dozen, or lots, too many for a startled brain to count. To say *I saw them* might mean exactly that or possibly less. *My friend saw them, he's unimpeachable.* Or maybe *I heard about it on pretty good authority.*

Scientific data are something else. They don't shimmer with poetic hyperbole and ambivalence. They are particulate, quantifiable, firm. Fastidiously gathered, rigorously sorted, they can reveal emergent meanings. This is why Mike Fay is walking across Central Africa with a little yellow notebook.

After two weeks of bushwhacking through Ebola's backyard, we emerge from the forest onto a red laterite road. Blinking against the sunlight, we find ourselves in a village called Minkoula, at which the dependable Tomo soon arrives with more supplies. Day 307 ends with us camped in a banana grove behind the house of a local official, flanked by a garbage dump and a gas-engine generator. The crew has been given an evening's furlough, and half of them have caught rides into Makokou to chase women and get drunk. By morning one of the Pygmies will be in jail, having expensively busted up a bar, and Fay will be facing a new round of political hassles, personnel crises, and minor ransom demands, a category of inescapable chores he finds far less agreeable than walking through swamp. But somehow he will get the crew moving again. He'll plunge away from the red road, diving back into the universe of green. Meanwhile he spends hours in his tent, collating the latest harvest of data on his laptop.

Within the past fourteen days, he informs me, we have stepped across 997 piles of elephant dung and not a single dung pile from a gorilla. We have heard zero gorilla chest-beat displays. We have seen zero sprigs of Marantaceae chewed by gorilla teeth and discarded. These are numbers representing as good a measure as now exists of the mystery of Minkébé.

Measuring that mystery is a crucial first step; solving it is another matter.

I make my departure along the laterite road and then by Cessna from the Makokou airstrip. The pilot who has come to chauffeur me is a young Frenchman named Nicolas Kozon, the same fellow who circled the Green Abyss at low altitude while Tomo tossed bombs of manioc and sardines to Fay and the others below. Now, as we rise from the runway, climb further, and point ourselves toward Libreville, the road and the villages disappear quickly, leaving Nicolas and me with a limitless vista of green. Below us, around us in all directions to the horizon, there is only forest canopy, and more canopy, magisterial and abstract.

Nicolas is both puzzled and amused by the epic daffiness of the Megatransect, and through our crackly headsets we discuss it. I describe the daily routine, the distances made, the swamps crossed, and what Fay faces from here onward. He'll visit the big waterfalls of the Ivindo River, I say, then turn westward. He'll cross the railroad line and two more roads, but otherwise he'll keep to the forest, following his plotted route, staying as far as possible from human settlements. He can do that all the way to the ocean. He'll cross the Lopé Reserve, yes, and then a big block of little-known terrain around the Massif du Chaillu. Another four months of walking, if all goes well. He's skinny but looks strong. He'll cross the Gamba complex of defunct hunting areas and faunal reserves along the coast, south of Port-Gentil, and break out onto the beach. He expects to get there in late November, I say.

With a flicker of smile, Nicolas asks: "And then will he swim to America?"

III. END OF THE LINE
November–December 2000

On the 453rd day of his punishing, obsessional, fifteen-month hike across the forests of Central Africa, J. Michael Fay stood on the east bank of a body of water, gazing west. It was not the Atlantic Ocean. That goal, the seacoast of southwestern Gabon, the finish line to his trek, was still twenty miles away. And now his path was blocked by a final obstruction, not the most daunting he'd faced but nonetheless serious: this blackwater sump, a zone of intermittently flooded forest converted to finger lake by the seasonal rains. Soaked leaf litter and other detritus had yielded the usual tannin-rich tea, and so the water's sleek surface was as dark as buffed ebony, punctuated sparsely by large trees, their roots and buttresses submerged. Submerged how deeply? Fay didn't know. Eighty yards out, the flooded forest gave way to a flooded thicket, a tangle of dense, scrubby vegetation with low branches and prop roots interlaced like mangroves, forming a barrier to vision and maybe to any imaginable mode of human passage. How far through the thicket to dry land? That also Fay didn't know.

"This is the moment of truth, I think," he said.

If it's only waist-deep, I said, with vapid good cheer, we could easily wade across.

"If it stays no deeper than *shoulder*," he corrected me, "we can make it." But he wasn't optimistic.

Fay took the machete of his point man, Emile Bebe, the young Baka Pygmy who had cut trail for him across hundreds of miles of Gabon. Slipping off his pack, wearing only his usual amphibious outfit (river sandals and river shorts), Fay waded out alone, probing the dark water ahead of him with a long stick. Bebe, two other walking companions—the photographer Nick Nichols and a videographer from National Geographic Television, Phil Allen—and I stood watching him go. Soon he was waist-deep, chest-deep, then armpit-deep, groping with his feet against sudden drops, seeking the shallowest route. Then there was just a little head and two skinny arms vanishing into the thicket. I climbed onto a woody loop of liana against the base of a tree, putting me six feet above the water and better positioned to listen, if not to see. I was concerned for him out there alone because of the crocodiles—not just *Crocodylous niloticus*, the fearsome Nile croc, but also a smaller species found hereabouts, *Osteolaemus tetraspis*, commonly known as the dwarf crocodile yet not to be taken too lightly. Of course my concern was futile, I realized, since from this distance, perched like a parrot on a trapeze, I couldn't give any timely help if a crocodile did grab him. I heard the whack of the machete. I heard fits of cursing, which alternated oddly with what sounded like bursts of demented song. We waited. He was gone for a half-hour, forty minutes, longer.

Meanwhile the rest of the traveling crew—two other Baka Pygmies and seven Bantu men, all carrying heavy packs of camp gear and scientific equipment and food, plus a middle-aged Gabonese forestry technician named Augustin Moungazi, whose role was to census trees—caught up and joined us at the water's edge. Where's the boss? they asked. Somewhere out there. The crewmen cast their eyes across the black lake with varying gradations of weariness and dread. Most of them had worked with Fay seven months now, since he had crossed into Gabon from the Republic of Congo, and they had been through such moments before. In the way they shrugged off their packs, uncricked their shoulders, inspected the route forward with leery scowls, they seemed to be saying: Oy, what manner of muddling travail gets us around *this* obstacle? It looked bad, but they had seen worse.

After nearly an hour I climbed down from my perch. Bebe smoked another cigarette. Nick aimed his Leica at anything remotely interesting. We swatted at filaria flies. We ate our crackers, nuts, and other piddling snacks representing lunch. We wondered silently whether Mike Fay would ever come back and, if not, how we'd find our way out of this forest without our mad leader. Then we heard shouts.

Fay had reached landfall beyond the thicket and returned just far enough to holler instructions. Mainly he was calling to the crewmen, in French, through the wattle of vegetation and the heavy equatorial air. Admittedly my French is lousy, Nick's and Phil's even worse, so we were befuddled; yet the Francophone crewmen appeared befuddled too. If we could just understand what Mike was saying, all of us, we would gladly comply. But to my ears he sounded like a bilious colonel of the French Foreign Legion screaming orders at new recruits through a mattress.

He had been right, in some sense, when he called it a moment of truth. Whereas Fay had come to study the forest, I had come to study Fay, and adversity is a great illuminator of true character. But then again, *truth?*—it's a quicksilver commodity, not so easily gathered as data. The moment was still unfolding, and so far there was more confusion than illumination. Did he want us to come or to stay? If we should come, then *how?* Should we cut logs and build a raft or just swim for it? The voice from the thicket seemed to convey almost nothing but purblind certainty and impatience. Was he mustering his troops for a final heroic lurch? Or, stressed by the long months of walking and the burden of forcing discipline on a group of freely hired men, by the nearness of the end, by his own ambivalence about reaching it, was he having a meltdown?

Days after this episode in the black lake, I would still be asking myself those questions. I would still be puzzling over the matter of J. Michael Fay and the complicated, provocative subject of leadership.

It was both the logic and the momentum of Fay's grand enterprise, which he had labeled the Megatransect, that had brought him and his entourage to this point of exigency on the 453rd morning. The logic was that he would walk a zigzag route from the northeastern corner of the Republic of Congo to the southwestern coast of Gabon, a distance of at least 1,200 miles, passing dead center through vast blocks of roadless and uninhabited forest, gathering data on vegetation, wildlife, and forest conditions as he went. The forest blocks, lying contiguous to one another, could be seen as gobbets of raw meat on Africa's last great kebab of tropical wilderness. Fay's route was to be the skewer.

The momentum derived from 452 days of footslog persistence, including many swamps mucked across and creeks forded, many resupply problems, many hungry nights, many nervous elephants with half a notion to make Fay himself a kebab, many hours of campfire laughter and bonhomie with the crew, many explosions of anger, many points at which it seemed almost impossible for Fay and his comrades to go on, after which they went on. Fay's logic insisted that this gargantuan transect be continuous and unbroken, both in space and in time. There had already occurred the one unavoidable gap, back in northwestern Congo just short of the Gabon border, when he departed from his plotted line to evacuate Mouko, who was verging on death from hepatitis. Although that short unwalked stretch—about eighteen miles, which he called the Mouko Gap—continued to nag Fay with a slight sense of incompleteness, he had put it behind him, marching on. By now his momentum included so many miles traversed (more like 2,000, in fact, than the 1,200 originally foreseen) and so many crises passed that it was unthinkable to be balked again, this time within twenty miles of the beach.

The logic of the enterprise had been laid out to the National Geographic Society (his main sponsor) and the Wildlife Conservation Society (his employer) in a forty-eight-page prospectus, with the forest blocks and his route sketched onto a multicolored map. The blocks as he had delineated them numbered thirteen, beginning with the Nouabalé-Ndoki block in northwestern Congo and ranging southwestward from there. Last in the chain was the Gamba block, a cluster of faunal reserves and defunct hunting areas along the Atlantic coast that are now being organized by the Gabonese government, with help from the World Wildlife Fund, into a complex of protected areas intended to preserve good habitat for elephants, hippos, dwarf crocodiles, and other sensitive species all the way to the beach.

Each of these blocks abuts another, and each is circumscribed by human impact (a road, a rail line, a string of villages along a river) but—this is the crucial part—virtually free of such impact at its interior. Although some armchair experts find it hard to believe, there *are* still sizable patches of African forest not currently occupied by human beings. Fay's concept was to travel by foot with a small support crew through these forest blocks and to measure in multiple dimensions the relationship between such absence of human impact and the ecological richness of the forest.

He described this data-gathering mission as a "reconnaissance survey," to distinguish it from the more formalized procedure known as the line-transect survey, wherein a field biologist walks and rewalks a short, straight path through the forest, gathering accretions of standardized data with each passage. Instead of cutting a ruler-straight corridor, Fay had elected to use a "path of least resistance" approach, letting the contours and obstacles of the landscape nudge him this way and that against his general compass bearing, and to make a single 1,200-mile walk instead of, say, 1,200 one-mile laps up and down a familiar snippet of trail. "The path of least resistance has the advantage of leaving the forest intact after passage, a significantly increased sample size because of increased speed, and considerably reduced observer fatigue," he had written in the prospectus. During my own time on the trail with him, totaling eight weeks divided into four stretches, I sometimes recollected the irony of that phrase, "the path of least resistance." It sounded lazily sybaritic, whereas here we were, clambering through still another tropical brier patch and then waddling across still another floodplain of sucking mud.

Now again on the morning of Day 453, as I squinted toward that thicket across the black lake, somewhere amid which Fay was hacking branches and yodeling orders, I had cause to wonder: This is the path of least resistance? Thank God we didn't come the hard way.

Like an unnerving omen of things to follow, Day 453 had begun with leeches. We had spent the night at Leech Pond Camp, thus dubbed by me (I named all the camps, for mnemonic purposes) when Fay returned from his evening bath and reported that ten leeches had gotten to him while he was rinsing. Leeches in moderation are no big deal, since they don't hurt and don't generally cause infection or carry disease. But the leeches that greeted us in the camp pond on the 453rd morning were beyond moderation. They swam up like schools of grunion and hooked their thirsty little maws to our ankles and calves, a half-dozen here, a half-dozen there, resisting slimily as we tried to pull them off. We had leeches under our sandal straps, leeches between our toes, leeches racing to every open sore. Good grief, what had they lived on before we arrived?

Hopping from foot to foot in the shallows, we deleeched ourselves while Bebe, also dancing and snatching at his feet between machete strokes, felled a small tree to bridge the pond's deeper trough. Then we tightroped across, deleeched again on solid ground, and went on.

Within a few minutes we heard monkeys jumping through the canopy. Fay did his usual trick, a whistling imitation of the crowned eagle, which provoked raucous alarm calls (*kaa-ko! kaa-ko!*) from the monkeys, allowing him to identify them: *Cercocebus torquatus torquatus*, the red-capped mangabey, locally known as the *kako*. He scribbled the exact time and the species name into his notebook, then took a five-minute sampling of their vocalizations on digital audio. Earlier he had mentioned that this mangabey, with its unmistakable flaming hairdo, was native only to forests near the Atlantic coast; farther inland, months ago, while crossing Congo and eastern Gabon, he had seen plenty of gray-cheeked mangabeys but none of the red-capped. Now here they were, offering a welcome signal that we had entered the coastal zone.

After an hour of easy walking along elephant trails, we found ourselves blocked by another dark pond. "Bad news, boys," said Fay. It looked as though the rainy-season waters were still up, he explained, which foreboded that there might be many such fingers of flooded forest between us and the coast. "If that's the case, we ain't gonna get through." But with a little scouting we found a fallen-tree bridge across the deep part and from there waded to dry land.

At the edge of the water stood another tree, a towering hulk with shaggy bark, a gracefully tilted trunk, and wide-reaching buttresses. Fay's routine called for noting every major tree along the route, so this one went into his little book: *Sacoglottis gabonensis*, 1.5 meters diameter near the base. Loggers generally ignore the species, he had said earlier, because its ropy, twisting trunks don't yield good lumber. The increasing abundance of *Sacoglottis gabonensis* was a further indicator that we were nearing the ocean. Still another was *Tieghemella africana*, a tree of high value both to timber companies and to elephants. Known commercially as *douka*, it grows to magisterial sizes—six feet in diameter and crowning out through the canopy—with straight, clean trunks, offering lovely wood for the sawmill. It also produces big green fruits, globular and heavy, each filled with sweet-smelling, pumpkiny orange pulp—not bad, but a little chalky to my taste. Elephants travel considerable distances to scarf douka fruits when they're ripe and falling, and the well-worn elephant trails we had been following seemed to run like traplines from one douka to another. Take away those mature, fruiting trees (by selective logging, for instance) and the local elephant population would lose part of its seasonal diet. But for now the grand old doukas were still here, showing evidence of recent attention (fresh elephant dung, gnaw marks in the bark), and so were the elephant trails. We hit another short stretch of good walking, then heard another group of monkeys.

This time, in response to the eagle whistle, there came a low, grunting chortle: *chooga-chooga-chooga-chooga-chooga.* Having heard it many times over the months, even I could recognize that as the alarm call of the gray-cheeked mangabey, *Lophocebus albigena,* another species dependent on fruiting trees. "It looks like the old gray-cheeks are gonna make it to the beach after all," Fay said. "That's cool. I was a little worried, 'cause we hadn't seen them for three or four days." The presence of *Lophocebus albigena,* overlapping here with its red-capped cousin, became another notebook entry. Then again we walked—westward, toward the beach—but only for five minutes, until the black lake stopped us cold.

The black lake: too wide to bridge and too long to bypass. According to Fay's map, it led northward into the Rembo Ngové floodplain, a riverine morass we didn't care to enter. So Fay had gone straight across, on his solitary probe, and was now out there somewhere in the thicket, shouting back instructions. Jean-Paul Ango, one of the youngest and strongest of the crewmen, took his machete to a modest-sized tree, which fell pointlessly into the water near shore. That can't be the idea, I thought.

Impatient with this muddle, I waded out along Fay's route to see if I could find the shallowly submerged ridge on which he seemed to have walked. Quickly I was neck-deep. So I decided to swim. Another crewman, Thony M'both, the man who told tales of Ebola at Mayibout 2, took the same notion at the same time, and we breaststroked across the black water on converging lines toward the thicket. Soon most of the crew had followed, some confidently, some reluctant to swim but more reluctant to be left behind. Strung out like a line of ducklings, they floundered variously with their waterproof packs, which were buoyant but too cumbersome to serve as water wings. Reaching the face of the thicket, Thony and I stopped. We treaded water. There seemed nowhere to go. I climbed up into the buttresses of a half-drowned tree, and one by one the others did likewise. In a neighboring tree I noticed Jacques Bosse, a big square-shouldered Bantu whom Fay had hired out of that gold-digging camp in northeastern Gabon. With a forceful yank, Jacques hoisted up his pack, to the outside of which was tied a large cook pot. He tossed back his head and muttered disgustedly to the sky that this was *no* kind of work for a man. We were stuck there, treed and frazzled like cats in a Mississippi flood, when Fay came out of the thicket and resumed command.

His first act was to holler sternly at Emmanuel Yeye, the shyest of the Pygmies, for letting his pack soak in the water rather than pulling it up. This gave way to a scathing harangue against the whole crew. Fay derided them for their fecklessness, their incompetence, their childishness and stupidity and insubordination. It was all in French, but what I missed in vocabulary I could gather from tone. It went on and on.

Nick and I had each witnessed earlier episodes of such castigation, going back to the first days of the Megatransect and Fay's Congo crew of Bambendjellé Pygmies. We had seen it after the walk through Minkébé, when some of the current crewmen got drunk and disorderly during their furlough at a resupply stop. We had seen it elsewhere. I had even begun to expect it (in my notes I called it, for shorthand, the Riot Act) as a calculated, self-conscious performance that Fay used periodically to restore discipline and focus. But this time both Nick and I felt he was going too far. Fay said blistering things of the sort that only a drill sergeant, an abusive father, or an especially caustic seventh-grade gym teacher might utter. He ranted and scorned. He recited the crew's failings. "*Ça me rend fou*," he growled repeatedly. "It makes me crazy."

Well, maybe so. At that moment, given our circumstances and the brave plunge these men had just taken, I thought that perhaps our brilliantly unorthodox Dr. Fay had indeed gone off his nut.

I was wrong. Later events and conversations with Fay, combined with what I knew of his personality and background, would convince me that this ultimate Riot Act tirade, as we all hung in trees above the black lake, was rational and carefully calibrated. Fay was stressed, yes, but still utterly in control. The deeper I scratched him, the more layers of ornery complexity and courageous bluntness I found. He wasn't always likable; sometimes he seemed piteously isolated; sometimes he seemed cynical and mechanistic about human relations; sometimes just too demanding and harsh. But in my final judgment, reached slowly, Fay is a formidable man with a strong sense both of mission and of fairness.

"Chaos breaks out very quickly and very easily," he would tell me days afterward, in the quiet of a tent pitched on a sandy hillock overlooking the Atlantic surf. "You've got to be a complete and utter hard-ass. And I don't enjoy being a hard-ass. I do not have some kind of sadistic element in my mind that makes me enjoy dominating people. But if you accepted that responsibility . . ." Thinking back over his fifteen months of risky travail, he dropped the second-person pronoun and spoke plainly. "Everything was my responsibility. Anyone who died on the Megatransect, it would have been my responsibility."

Mouko had nearly died of hepatitis, and it was Fay who nursed him until evacuation was possible. A crewman named Roger had almost drowned, tangled in his pack straps at a river crossing, when he larkishly flouted Fay's instructions. There had been several other close calls in water, and several other medical emergencies. "I take that very seriously," Fay said. "And I take the data collection very seriously." All along, he explained, he'd had three overriding goals: to finish the entire walk as originally conceived, to maintain an unbroken regimen of data-gathering, and to get everyone through the experience alive. Democracy on the trail and his own popularity on any given day were not even secondary concerns.

The data would eventually be collated, cross-referenced, elaborately crunched and analyzed during the months of his follow-up work. Which forests seem to be richest in gorillas? How quickly do elephants recolonize an area where elephants have been poached? What's the linkage between logging roads and the presence or absence of duikers, forest hogs, cercopithecine monkeys? Everywhere, every which way, he wanted to ask and to answer: What are the correlations? He hoped that question would lead to another: What are the implications for wise management? Fay would write a report or a book, maybe both, and then also make it all available through a Web site.

"On this Web site people are going to see very clear patterns," he vowed as we sat above the beach. "Nobody is going to be able to deny that there is something there." Referencing one slice of data to another would in some cases yield high statistical correlation, and observers (so he imagined, supplying their words) would say, "Wow! Look at this, man. Douka, elephant: correlation, point nine." He meant that the degree of congruence between douka trees and elephant trails might be .9, of a possible 1.0. "That's pretty cool." Fay hoped, anyway, that observers would find such a relationship and offer such a response. "Just seeing those patterns is going to make people realize that this is a viable methodology."

But before any such epiphanies could happen, he needed the data. They had to be continuous. They had to be rigorous. Toward that end, his organizational model for the Megatransect was unabashedly autocratic. During one milder fit of annoyance, provoked by a food shortage after some crewmen had evidently jettisoned provisions in order to lighten their packs, I heard Fay tell the men that if this were a military operation, they would all be in prison. "It's very much like a military operation," he said to me now. "I am the commander in chief on the Megatransect." That might sound "radical" to some ears, he acknowledged (or maybe "offensive" or "retrograde" are the words you want there, I thought), but to anyone who had shared the many months of daily effort and frequent peril, it would make perfect sense. There could be only one leader giving orders, and those orders had to be followed without malingering or debate, or else the whole effort would unravel and the three goals wouldn't be met.

Where did this military style come from? Fay is too young to have experienced Vietnam or the draft, too old to have signed on for the first Iraq war, and has never served in any branch of the armed forces. It's hard to imagine how he ever could have. Three or four months of basic training and regimentation would no doubt have aggravated his own insubordinate tendencies to the point of court-martial or discharge. But the lore of certain military operations intrigued him—Vietnam particularly, maybe because it was a jungle war and he's a jungle guy. The American fiasco there, I once heard him argue, reflected the plain fact that American troops weren't at home in the ecosystem. (Undoubtedly that had been part of it, along with several millennia's worth of cultural differences and political history, plus a few other factors.) During my earlier visits to the line of march, I had brought him some of the better Vietnam memoirs for trail reading—Doug Peacock's *Grizzly Years*, Michael Herr's *Dispatches*—which seemed to engross him for a few hours at night after his data-entry chores. He occasionally mentioned that if he weren't an ecologist, he might be tempted to find work as a war photographer. And he was fascinated by the Lewis and Clark expedition, which besides being an exploratory trek was a mission, under full discipline, of the United States Army.

Back at the start of the Megatransect, in the disheveled little library of his cabin in Bomassa, the research camp in northeastern Congo that had served Fay in recent years as a base, I had found his own dog-eared and heavily marked copy of *Undaunted Courage*, Stephen Ambrose's account of the life and character of Meriwether Lewis, as revealed most gloriously during his journey with Clark. That journey was, of course, America's own first and greatest Megatransect. One passage in the Ambrose book, completely underlined by Fay, caught my eye: "Two years of study under Thomas Jefferson, followed by his crash course in Philadelphia, had made Lewis into exactly what Jefferson had hoped for in an explorer—a botanist with a good sense of what was known and what was unknown [and] a working vocabulary for description of flora and fauna, a mapmaker who could use celestial instruments properly, a scientist with keen powers of observation, all combined in a woodsman and an officer who could lead a party to the Pacific." A botanist, a woodsman, a leader. Reading that, Fay must have felt some tingle of identification.

Never mind the sad fact that Meriwether Lewis, addled by acclaim and alcohol after his big success, eventually killed himself. Fortunately for Fay, the parallel between him and Lewis isn't really so close. Lewis stepped into a mission that had been dreamed up by President Jefferson, whereas Fay himself, no one else, concocted this one. Lewis and Clark's enterprise was premised upon the goals of commercial exploitation and easy travel for traders, whereas Fay's Megatransect has a drastically different goal: protecting big areas of rich forest from reductive human impact. Fay has had a better scientific education than Meriwether Lewis, and unlike Lewis, he seems not very susceptible to booze or self-doubt. Another advantage is that whereas Lewis headed off into a difficult sort of landscape he'd never seen before, Fay had twenty years' experience in various Central African forests.

He knew the ecosystems from bottom to top—from the plants to the elephants and the gorillas. Equally important, he knew how to walk through this world. Beginning in the late 1980s, when he did his doctoral fieldwork on lowland gorillas in the Central African Republic, tracking them through the forest with his Pygmy mentor, Mbutu Clement, Fay developed the habit of making long, restless explorations by foot. The little Suzuki trail bike that he had used during his Peace Corps days, up in the savanna country near the border with Chad, was no longer useful and no longer necessary. He discovered that by adapting his body and his outfit (river sandals, one pair of shorts, and no shirt, since bare skin is more easily washed and dried than clothing) to local conditions, he could cross flooded forests, streams, boggy clearings, and swamps that most other people considered impassable. He also learned that he could walk into a village or town anywhere in Central Africa and, within a day or two, hire a crew of men who were glad for the work of carrying bags and making camps. Employment was scarce, and he paid better than most. He learned how many men were required for transporting this much scientific equipment, that many tents, and enough food to sustain them all for, say, twenty or twenty-five days between points of resupply. By trial and error he developed a style of personnel management that worked.

One element of that style was his imperious sense of command. Another was that he never asked anyone to accept discomfort or risk that he wouldn't accept himself. The historian Plutarch, in his life of the Roman general Marius, wrote that "there is nothing a Roman soldier enjoys more than the sight of his commanding officer openly eating the same bread as him, or lying on a plain straw mattress, or lending a hand to dig a ditch or raise a palisade. What they admire in a leader is the willingness to share their danger and hardship, rather than the ability to win them honour and wealth, and they are more fond of officers who are prepared to make efforts alongside them than they are of those who let them take things easy." In Fay's case, it was manioc and salted fish, not bread; a roll-out pad on the forest floor, not a straw mattress; and a machete-cut corridor through a blackwater thicket in lieu of a raised palisade.

When I asked him later about his blowup at the black lake, he conceded that "it certainly looked like I was pissed off, there's no doubt about it." And yet he hadn't been, he said. It was just another bit of tactical histrionics. From his perspective (though he was too discreet to say so), I had exacerbated the confusion myself when Thony and I triggered the group swim. He had intended to proceed methodically, but my impatience foiled that. "I was simply taking chaos and putting order into it. And the only way to do that is to say, at the top of your lungs, 'Everybody *stop!* Everyone who is here present, *stop!* Do not move. Do not breathe. Stop. And I'm going to tell you what to do.'"

Fair enough, though as the moment had unfolded, I didn't wait to be told. I swam back to the east side of the lake, found my own waterproof pack where I had left it, double-checked its seal for the sake of my notebook and binoculars, and swam out again to the thicket. By the time I got there, nudging the pack ahead of me like a water polo ball, the others had begun moving down Fay's hacked-out corridor. The water here seemed to be eight or ten feet deep. I fell in behind Sophiano Etouck, the most stalwart of the crewmen, and Nick, who was managing somehow to dogpaddle along with his pack on his back and his Leica to his face like a snorkel mask.

God love him, Nick even now was shooting. Sophiano led the way, swimming with his right arm and wielding a machete with his left. Every few yards he rose high in the water to whack a limb out of our path, then sank away beneath a boil of bubbles. When Sophiano first went under and stayed under, Nick and I both worried that he had tangled himself in some vegetation; then, exuberant as an otter, he exploded back up to take another swing. I followed him for fifty yards through this watery tunnel of limbs and roots, a passable route that Fay had opened during his missing hour. Finally the thicket cleared, the water shallowed suddenly, and we climbed up a high bank onto firm ground.

While Nick and Phil examined their cameras for damage and their bodies for leeches, I dropped my pack and went back in the water to see if I could help with another load. After swimming down one blind alley, I found the tunnel again and retraced it to the east edge of the thicket. Fay was there, still perched in a tree, having meanwhile swum the lake to retrieve his own pack.

Now he was shepherding along the last of the crew. He knew from experience which of the men were steady swimmers and which needed assistance. He was giving instructions, but the strident moment had passed. In fact, he seemed subdued. I took the pack of Augustin, the botanist, who preferred climbing through the thicket to swimming under it, and Fay came behind all of us as sweeper. He even brought my sleeping pad, which had gotten unpacked during some emergency reshuffling of the loads and been temporarily stowed in a tree. He handed it back to me dry.

By 12:40 P.M. we stood on the west bank, wringing out our shirts (except for Fay, still shirtless), checking our packs for leakage, basking in the sunshine—rare sunshine!—that blessed us there through a canopy gap. Flush with nervous relief, we joked and relaxed. We were pleased with ourselves for having wiggled through what might be, we hoped, the last of the dire obstacles. Emmanuel lifted a sodden ten-pound bag of rice from his pack, letting the pale milky water drain out. Nick labeled rolls of film. Sophiano had a smoke. Fay, head down, quietly wrote in his yellow waterproof notebook.

And then, without comment, without any speech of further

remonstrance, let alone congratulations, Fay detached himself crisply from our breezy mood. He glanced at his wrist compass. He turned toward the forest and stabbed out his arm, giving the usual signal to Bebe: *That* way. Dutifully, Bebe stepped out and began cutting trail. Fay walked.

Snatching up my pack, holstering my notebook, I followed. I was startled by his brusqueness, but I wanted to stay at Fay's heels. Maybe in the aftermath he would loosen up and commit a personal revelation. Maybe he'd put his outburst in context. Or maybe he'd just encounter something interesting—a Gaboon viper, a gorilla, a dwarf crocodile—that I'd hate to miss. The rest of the party were left behind to think what they might think, to feel what they might feel, to gather themselves at their own pace.

At 1:11 P.M. on Day 453, Fay paused to record the next datum: elephant dung, old. Then again, without speaking, he walked. A hard man, a savvy leader, a flouter of pieties, a solitary soul, a conscientious scientist, a fierce partisan of tropical forest, a keen judge of human limits, he had work to do—not much work remaining now, but some. He couldn't celebrate yet. He was still three days from the beach.

A Passion for Order

SPRINGTIME COMES LATE IN SWEDEN. So it was still springtime, on May 23, 1707, when a son was born to the wife of the curate of a small Swedish village called Stenbrohult. The season was raw, the ground was wet, the trees were in leaf but not yet flowering as the baby arrived, raw and wet himself. The child's father, Nils Linnaeus, was an amateur botanist and an avid gardener as well as a Lutheran priest, who had concocted his own surname (a bureaucratic necessity for university enrollment, replacing his traditional patronymic, son-of-Ingemar) from the Swedish word *lind*, meaning linden tree. Nils Linnaeus loved plants. The child's mother, a rector's daughter, had barely turned eighteen. They christened the boy Carl, and as the story comes down, filtered through mythic retrospection flattering a man who became the world's preeminent botanist, they decorated his cradle with flowers.

When he was cranky as a toddler, they put a flower in his hand, which calmed him. Or anyway, again, that's what later testimony claims. Flowers were his first keen interest and his point of entrance to appreciating beauty and diversity in nature. He seems even to have sensed at an early age that they were more than just beautiful and diverse—that they also encoded some sort of meaning.

He grew quickly into a boy fascinated not just by flowers and by the plants that produce them but by the *names* of those plants. He pestered his father for identifications of wildflowers brought home from the meadows around Stenbrohult. "But he was still only a child," according to one account, "and often forgot them." His father, reaching a point of impatience, scolded little Carl, "saying that he would not tell him any more names if he continued to forget them. After that, the boy gave his whole mind to remembering them, so that he might not be deprived of his greatest pleasure." His greatest pleasure: connecting labels with living creatures and memorizing them. This is the sort of detail, like Rosebud the sled, that seems too perfectly portentous for real history as opposed to screen drama or hagiography; still, it might just be true. Names and their storage in memory, along with the packets of information they reference, are abiding themes of his scientific maturity.

But to understand the huge renown he enjoyed during his lifetime, and his lasting significance, you need to recognize that Carl Linnaeus wasn't simply a great botanist and a prolific deviser and memorizer of names. He was something more modern: an information architect.

If you read a thumbnail biography, in an encyclopedia or on a Web site, you're liable to be told that Carl Linnaeus was "the father of taxonomy"—that is, of biological classification—or that he created the Latin binomial system of naming species, still used today. Those statements are roughly accurate, but they don't convey what made the man so important to biology during his era and afterward, nor why we should honor him three centuries later. You might read that he coined the name *Homo sapiens* for our own species and placed us, daringly, within a category of mammals that included also monkeys and apes. That's true too, but somewhat misleading. Linnaeus was no evolutionist. He helped pave the way toward evolutionary thinking, but he never traveled that bumpy route himself. On the contrary, he heartily embraced the creationist view of biological origins, as formalized in a body of thought called natural theology, which stipulated that studying nature reveals evidence for the creative powers and mysterious orderliness of God. Linnaeus wasn't such a devout man, though, that he sought nothing *but* godliness in the material world. His intellect and his spirit were more scientific than pious. Here's what makes him a hero for our time, and for all times: He treasured the diversity of nature for its own sake, not just for its theological edification, and he hungered to embrace every possible bit of it within his own mind. He wanted to know what lived. He believed that mankind should discover, name, count, understand, and appreciate every kind of creature on Earth.

In order to assemble all that knowledge, two things were required: tireless and acute observation, and a system.

In the spring of 1732, just before his twenty-fifth birthday, Linnaeus set off on an expedition through Lapland, the northern region of the Swedish kingdom that was then one of Europe's great remaining wilderness areas, inhabited by a sparse population of the Sami people (known to outsiders as Lapps), who lived as herders of reindeer. He carried an extra shirt, a pen case, a magnifying glass, a sheaf of paper for pressing plant specimens, a gauze veil to protect him from midges, a short sword, a comb, and a few other pieces of vital gear. Over the next five months he traveled about 3,000 miles, by horseback and foot and boat, making collections and taking notes as he went. He was interested in everything— birds, insects, fishes, geology, the customs and technology of the Sami—but especially in plants. He made drawings in his journal, some of which were crude sketches, some of which (again, those of plants) were delicate and lovingly precise. The Lapland journey was a formative adventure for Linnaeus, just as other journeys would be for other young naturalists, such as von Humboldt in South America, Darwin on the *Beagle*, and T. H. Huxley aboard H.M.S. *Rattlesnake*. Eventually he produced a book, *Flora Lapponica*, describing the botanical data he had gathered, but he left his travel journal unpublished, to be discovered by scholars after his death.

He went abroad in 1735 to advance his career prospects. He spent three years on the continent, mostly in Holland, taking a medical doctor's degree quickly, then turning back to plants. It wasn't a stretch to combine both activities, since botany in that era was closely related to medicine through the pharmaceutical uses of vegetation, and biology per se wasn't yet a profession. Most botanists (most zoologists too) made their living as doctors or clergymen. Linnaeus found temporary work with a rich man named George Clifford, a director of the Dutch East India Company, as botanical curator and house physician at Clifford's country estate near Haarlem. Clifford's connection to an international trading company had helped him gather a showcase garden and a zoo full of exotic animals. Linnaeus's work there led to another book, a descriptive catalogue of Clifford's holdings titled *Hortus Cliffortianus* and gorgeously illustrated by a young botanical artist named Georg Dionysius Ehret. Although they became lifelong friends, Ehret later recalled the Linnaeus of these years as a self-aggrandizing opportunist. By any account, he was full of energy and ambitions, full of ideas and opinions, impatient with elder authority, not reluctant to borrow good material from others, hungry for success as well as for deeper knowledge. He had great facility as a writer, both in Swedish and in the simplified Latin then standard for international scientific discourse. He had social facility too. Confident to the point of arrogance but charming enough to compensate, he proved good at making friends, finding sponsors, and cultivating powerful contacts. During the three years abroad he published eight books—an amazing spurt of productivity, partly explained by the fact that he had left Sweden carrying some manuscripts written earlier. One of those manuscripts became *Systema Naturae*, now considered the founding text of modern taxonomy.

Linnaeus wasn't the first naturalist to try to roster and systematize nature. His predecessors included Aristotle (who had classified animals as "bloodless" and "blooded"), Leonhart Fuchs in the sixteenth century (who described five hundred genera of plants, listing them in alphabetical order), the Englishman John Ray (whose *Historia Plantarum*, published in 1686, helped define the species concept), and the French botanist Joseph Pitton de Tournefort, contemporary with Ray, who sorted the plant world into roughly seven hundred genera, based on the appearance of their flowers, their fruit, and their other anatomical parts.

Linnaeus emerged from this tradition and went beyond it. His

Systema Naturae, as published in 1735, was a unique and peculiar thing: a folio volume of barely more than a dozen pages, in which he outlined a classification system for all members of what he considered the three kingdoms of nature—plants, animals, and minerals. The section on minerals was the least interesting and influential. What really mattered were his views on the kingdoms of *life*.

His treatment of animals, presented on one double-page spread, was organized into six major columns, each topped with a name for one of his classes: Quadrupedia, Aves, Amphibia, Pisces, Insecta, Vermes. Quadrupedia was divided into several four-limbed orders, including Anthropomorpha (mainly primates), Ferae (such as canids, felids, bears, possums), Pecora (including deer, antelope, cattle), and others. His Amphibia encompassed reptiles as well as amphibians, and his Vermes was a catchall group, containing not just worms and leeches and flukes but also slugs, sea cucumbers, octopuses, starfish, barnacles, limpets, corals, and other sea animals. He divided each order further, into genera (some with recognizable names such as Leo, Ursus, Hippopotamus, and Homo), and each genus into species. Apart from the six classes, Linnaeus also gave half a column to what he called Paradoxa, a wild-card group of chimerical or simply befuddling creatures such as the unicorn, the phoenix, the dragon, the satyr, and a certain giant tadpole (now known as *Pseudis paradoxa*) that, weirdly, shrinks during metamorphosis into a much smaller frog. Across the top of the chart ran large letters: CAROLI LINNAEI REGNUM ANIMALE. It was a provisional effort, grand in scope, integrated, but not especially original, to make sense of faunal diversity based on what was known and believed at the time. Then again, animals weren't his specialty.

Plants were. His classification of the vegetable kingdom was more innovative, more informed, and more orderly. It became known as the "sexual system" because he recognized that flowers are sexual structures and he used their male and female organs—their stamens and pistils—to characterize his groups. He defined twenty-three classes, into which he placed all the flowering plants (with a twenty-fourth class for cryptogams, those that don't flower),

NATURAL ACTS | 284

based on the number, size, and arrangement of their stamens. Then he broke each class into orders, based on their pistils. To the classes he gave names such as Monandria, Diandria, Triandria (meaning one husband, two husbands, three husbands) and within each he gave ordinal names such as Monogynia, Digynia, Trigynia, thereby evoking all sorts of scandalous ménage (a plant of the Monogynia order within the Tetrandria class: one wife with four husbands) that caused lewd smirks and disapproving scowls among some of his contemporaries. Linnaeus himself seems to have enjoyed the sexy subtext. And it didn't prevent his botanical schema from becoming the accepted system of plant classification throughout Europe.

The artist Georg Ehret again helped popularize his ideas by producing a handsome *tabella*, a poster, illustrating the diagnostic features for Linnaeus's twenty-four classes. The tabella sold well and earned Ehret some gulden. Linnaeus, always stingy about sharing credit, included Ehret's drawing without acknowledgment in one of his later books. But he wouldn't forget his old pal, and evidence left after his death—we'll come to it—suggests that he valued Ehret's botanical vision as he valued few aside from his own.

After returning to Sweden, becoming a husband and a father and a professor at Uppsala University, Linnaeus continued to churn out books. He published revised and expanded editions of *Systema Naturae* as well as strictly botanical volumes such as *Flora Suecica* ("Swedish Flora") in 1745, *Philosophia Botanica* (1751), and *Species Plantarum* (1753). *Philosophia Botanica* is a compendium of terse, numbered postulates in which he lays out his botanical philosophy. For instance: "The foundation of botany is two-fold, arrangement and nomenclature." Arrangement of plants into rational categories and subcategories is crucial for three reasons: because there are so many kinds (and more every year, during the great age of discovery in which Linnaeus lived), because much is known about many of those kinds, and because classification makes that knowledge accessible. Alphabetical listing may have worked well enough with five hundred plant species, but as the count rose into thousands, it didn't serve.

There was also a deeper purpose, for Linnaeus, to this enterprise. Find the "natural method" of arranging plants into groups and you would have discovered God's own secret logic of biological creation, just as Isaac Newton had discovered God's physical mathematics. Linnaeus knew that he hadn't achieved that, not even with his twenty-four-class sexual system, which was convenient but artificial. He couldn't see, couldn't imagine, that the most natural classification of species reflects their degree of relatedness based on descent from common ancestry. But his passion for order—for seeking a *natural* order—did move taxonomy toward the insights later delivered by Charles Darwin.

As for nomenclature, it contributes to the same purpose. "If you do not know the names of things, the knowledge of them is lost too," he wrote in *Philosophia Botanica*. Naming species, like arranging them, became increasingly problematic as more and more were discovered; the old-fashioned method, linking long chains of adjectives and references into fully descriptive labels, grew unwieldy. In *Species Plantarum* he established the Latin binomial system for naming plants, and then in the tenth edition of *Systema Naturae*, published in 1758–59 as two fat volumes, he extended it to all species, both plant and animal. A pondweed clumsily known as *Potamogeton caule compresso, folio Graminis canini . . . et cetera* became *Potamogeton compressum*. We became *Homo sapiens*.

His life back in Uppsala entailed more than authorship. He was a wonderful teacher with a vivid speaking style, clear and witty, and a terrific memory for facts. His lectures packed the hall, his private tutoring earned him extra money, and he made botany both empirical and fun by leading big festive field trips into the countryside on summer Saturdays, complete with picnic lunches, banners and kettledrums, and a bugle sounding whenever someone found a rare plant. He had the instincts of an impresario. But he was also quietly effective in mentoring the most talented and serious of his students; almost two dozen of them became professors themselves, and more than a dozen went off on adventuresome natural-history explorations around the world, sending data and specimens faithfully back to the old man. With his typically sublime absence of modesty, he called those travelers "the apostles." In 1761 the government ennobled him, whereupon he upgraded his linden-tree name to von Linné. By then he was the most famous naturalist in Europe.

His wife sternly guarded their pirvacy, and his son became only a middling botanist, but his teaching role delivered rich satisfactions, and he had an abundance of brilliant intellectual offspring. Despite the limitations of his language skills (he may have known some Dutch and German but did all his writing in Swedish and Latin) and of his geographical experience (he never left Sweden again), he became a global encyclopedist of flora and fauna through his written correspondence with naturalists all over the world and through the information he received from the apostles, such as Daniel Solander (who sailed on Cook's first voyage), Pehr Kalm (in North America), Fredrik Hasselqvist (the Middle East), Pehr Löfling (Spain, Venezuela), and Anders Sparrman (China, South Africa, then Cook's second voyage). Some of them, such as Löfling and Hasselqvist, never got back alive, but at least Hasselqvist's manuscripts found their way home, thanks to a ransom (he died in debt) paid on behalf of Linnaeus by the queen. Linnaeus himself had no appetite for the rigors and climate of the tropics, though he was voraciously curious about tropical plant diversity. Let the young men gather the information; he would systematize it.

On a recent afternoon in Uppsala, I discussed this manipulative, homebody aspect with Professor Carl-Olof Jacobson, a retired zoologist who serves as chairman of the Swedish Linnaeus Society. No, Linnaeus didn't want to travel, Professor Jacobson told me. "What he wanted to be was a spider in the net."

The center point of that net, that vast web of scientific silk, was in and around Uppsala—including the university, its splendid botanical garden, and a small farm known as Hammarby, about five miles outside the city. Linnaeus bought Hammarby and built a large, simple house there to be his summer retreat. It might have served also as his retirement home, though he never retired. Near the west wall, visible from his bedroom window, he planted a cherry tree. On a rocky hill just behind, he added a small square cabin, lacking a stove and therefore fireproof, as repository for his most precious specimens. Each autumn he moved back into town, where the living was less austere. He grew feeble and ill, then suffered a seizure after one last escape to the countryside, strictly against doctor's orders, and died on January 10, 1778. They buried him beneath the stone floor of Uppsala's cathedral, the Westminster Abbey of Sweden.

Six years later, following Linnaeus's posthumous instructions, his widow sold his library, his manuscripts, and most of his collections to a buyer who would care for them well. That buyer, a young Englishman named James Edward Smith, founded a scientific society to receive the treasures and called it the Linnean Society (its spelling derived not from Linnaeus but from the noble version, von Linné) of London, where they lie protected today in a basement vault, awaiting scholars to come visit them. Knowledge, the man always argued, is meant to be communicated and used.

But his country home, Hammarby, remained in the family for a century and then was bought by the Swedish state to be made a museum. Although his house near the university in Uppsala has also been saved, and lately restored, Hammarby conveys a more vivid sense of his character, his foibles, his loneliest joys. Inside the old farmhouse, overlooking muddy crop fields, his collection of walking sticks is on display. So is the red skullcap he often wore over his short-cropped hair, in lieu of a formal wig. There are portraits of his four daughters, his son, and his pet monkey, in no particular order of fondness. His wife and he kept separate bedrooms at opposite ends of the second floor. His bedroom is tucked away, accessible only through another room, which functioned as his study.

The bedroom, preserved much as he left it, contains a small curtained bed of the sort known in Sweden as a *himmelssäng*, a "bed of heaven." Against the west wall is a wooden desk and, above it, a window. The walls are covered with flowers.

That is, they are wallpapered wildly from floor to ceiling with large floral images cut from books. The plants are robust, exuberant, some of them garish, some elegant, all suggesting fecundity and fruition: pineapple, banana, magnolia, lily, cactus, papaya, frangipani, and others. Many of these hand-colored engravings came from paintings by his old friend Georg Dionysius Ehret. Rare and magnificent, they would be collectables in their own right, even absent the association with Linnaeus. But, once bright and crisp, they are now faded, smeary, streaked with the punishments of moisture and time. On the day I visited, accompanied by a botanical curator named Karin Martinsson, still another damp January chill hung in the air.

Linnaeus was warned that such damage would occur, but evidently he didn't care. He wanted the pictures around him. Never mind if they decayed. So what? His own body was doing that too.

Even now these antique prints could be peeled carefully off, Martinsson told me, and preserved under better conditions. But that's not going to happen. "Taking them down from the walls," she said, "would be like ripping the heart out of Hammarby." Left as is, the heart of the house reflects the heart of its original owner: full of plants. The pilgrims who visit this room during the tercentenary year—presumably there will be many, from around the world—can look at that improvised wallpaper and sense an important truth about the life's work of Carl Linnaeus. It wasn't just about knowledge. It was about knowledge and love.

Citizen Wiley

THIS IS A STORY OF MONTANA but without horses and cows. The romance of place is not its subject. It will contain no comparisons to Paris in the twenties or Tangier during the era of Paul Bowles. A river doesn't run through it. The horses and cows belong to other people's Montana stories, not mine; place itself, not the romance of place, is what I want to evoke; and though rivers have in fact been extremely important to my own life in Montana, they aren't the point here. Most of the state's landscape consists of mountains and wide valleys and sage-covered hillsides and high plains transected by long empty stretches of two-lane blacktop, true enough, but this story takes place in town. Its moral, if it has one, is probably "You never know where you're going until you've been there quite a while." Its central character is a large gray dog of indeterminate lineage and advanced age. But he doesn't make his entrance until later.

Thirty-three years ago I arrived in Montana in a Volkswagen bus, into which was crammed most of what I owned, or anyway most of what seemed to me valuable: a fishing rod, a Smith-Corona typewriter, and many paperback books. I settled in a town just over the Idaho border, where three rivers converged, and found myself an apartment. It was a college town, eclectic and libertine compared to other parts of Montana at the time. For instance, you could buy a cup of espresso. I thought this might be an agreeable place to live, given the things that I cared about (trout, mountains, snow) and the other things (cities, Ohio, august universities) that I hankered to escape; and I assumed that it should be as good as anywhere, or almost anywhere (*no*, one friend told me: *New York*) for a fellow who wanted to become a writer. I had already published one book, but I hadn't learned the first crucial lesson about a literary vocation: *Don't quit your day job, not yet!* I'd never even had a day job. Now I discovered I needed one.

Montana wore on me well. The winters were long and severe, which suited the Norwegian side of my genetic predisposition; the people were friendly and trustful; children would look a grown man in the eye, a stranger, and say hello; shop clerks and workingmen harbored prideful affection toward the state in which they happened (or chose) to live, and that mere geographical dimension of contentment seemed to help make their lives rather more happy than hard. Very soon, I felt it too.

That's not to say I fell into a life of ease and frolic. There were some years of groping, dues-paying, wage-laboring, writing and writing but not publishing, frustration, rethinking, adjustment. There were relocations from one part of the state to another as opportunity, necessity, or whimsy decreed. There was one job, in Butte, that involved reporting to an office. When that ended abruptly, I decided henceforth to scratch out my living as a freelancer. I moved again, this time to a fishing village on another river, an enclave for getaway artists and remittance men. I adopted a cat. I got married. I spent a brief exile in Arizona and returned. Then, in 1984, I fetched up in a town called Bozeman, population about 30,000, in a broad honest valley between the Bridger Mountains and the Gallatins. A year later I settled onto the small patch of land where I remain anchored today: a narrow lot in a tree-shaded neighborhood just a few blocks south of Bozeman's Main Street.

My life here is quiet. In winters I ski, shovel snow, and play hockey with a town-league team composed of wonderful men I wouldn't otherwise know—a roofing contractor, a gun dealer, an insurance adjustor, a surveyor, a truck driver, a software designer, a civil engineer, and others. In summers I race bicycles with retired but dauntingly fit academics. For part of each year I travel to faraway places, on magazine assignments or for book research; then I come home and sit in my office and write. That's a change from earlier years, when I intended to be a novelist creating books by sheer force of imagination, and when I had no money for plane tickets anyway.

Other things too have changed, more cataclysmically, since I came to Bozeman. After thirteen years of feeling happy but increasingly cramped in a ramshackle bungalow, my wife and I dismantled the original house, board by board, with skilled help, and built a new one on the same patch of dirt. It's a handsome structure of recycled wood, craftsman style, with a turret, which cost moderately in dollars and egregiously in other ways. During the course of that project, the marriage became dismantled too. My ex-wife has gone on to a new career, a new situation, in which I'm glad to see her thriving, and I've held on to the house. Or I should say, it has held me. Much of its wooden tissue, old two-by-eights of long-grain fir and floorboards of oak, is the same wood I've lived amid since 1985, now reconfigured and repositioned unrecognizably, like flesh from a thigh grafted onto a burn victim's face. That feels weird but good. The mountain ash tree growing snug to the new house is the same mountain ash that hugged the old house. Its root structure survived the excavation. The elm in front is also the same. The original cat, long since buried in back beneath the old apple tree, lies there still—her remnant molecules, anyway—now shaded by a new apple tree. The tumbledown shed, where I store a lawnmower and two dusty kayaks, is likewise unchanged. In twenty-one years, I've never quite gotten it cleaned and organized. And many of those tattered paperbacks that came with me in the VW bus—including the two-volume *War and Peace* in a slipcase and the *Catch*-22 inscribed by Joe Heller himself—are still here on the shelves.

So much for doddering continuity. I can also report cheerful news of vibrant renewal: Three years ago I married a woman who carried her own style of joy, her own large and irrepressible heart, into this half-empty house. I won't try to describe all she has added to my life and to the place where I live it, since that too isn't the point, and this essay is short. I'll mention just one crucial thing: She brought with her a large gray dog.

His name was Wiley. She had adopted him years before, when

he was a roguish young stray who crashed an Earth First! picnic, made himself popular, and was rewarded with burgers and cake by the handful. For a while he ran with a rugby team. He was so badly behaved that several times in the early years he nearly lost his life to irate ranchers incensed at this overgrown cur who was stampeding their livestock. Although his parentage was unknown, from the look of his burly build, his fluffy tail, his thick gray fur, his floppy right ear, and his burning orange eyes, he seemed to be the love child of a Great Pyrenees and a wolf. When pressed on the subject, though, his mistress usually passed him off as a "malamute–golden retriever cross," not wanting to add to the misguided vogue for wolf-hybrid pets. (He did show some golden blond highlights, on which basis I later framed my own answer to the recurrent question about lineage. He's a *malomino*, I'd say: half malamute, half palomino. Have I mentioned that he was large?) He entered my life when she did, almost five years ago, and like her, he changed it.

I had never been what you'd call a dog person. That's an understatement. I had felt, spoken, and written some scurrilous things about the canine race. But this Wiley was different from any dog I'd ever encountered: more joyous, more handsome, more dignified and laconic (very disinclined to bark), more confident and commanding, yet peaceable. Children could tug at him with impunity. He'd never start a dogfight, but he'd gladly finish one. I'm utterly biased, of course, because he became my stepdog, and then simply "my dog" (insofar as one can say "my dog" about any half-feral foundling) as well as my wife's. Notwithstanding my bias, other people seemed to recognize Wiley's unusual charms. He had a face that made folks happy from merely gazing into it. He enjoyed trotting up to say hello. He worked crowds like a candidate for Congress. People around town knew him by name, I noticed, who didn't know me and whom I didn't know. He retained his enthusiasm for crashing parties, especially those at which steaks were left unattended on low tables. He was horribly unqualified as a watchdog because he loved everyone; Saddam Hussein or Richard Speck could have snuck in our back door, with or without raw hamburger, and been met with a lick on the hand. Wiley would have done better as a greeter in Reno.

He slept at night on the floor of my office. During cocktail hour he slept on the living room carpet, between my wife's chair and mine. He slept by day in the backyard, and if the weather was wintry, all the better. His fur was so thick he seemed incapable of feeling chilled, and he liked especially to lie out there during a snowstorm, nose to tail, letting the flakes coat him like rime.

Soon after he joined the household, Wiley and I began to walk the neighborhood together in a way I'd never walked it alone. My old solitary morning trudges, usually before dawn, had looped eastward from the house and involved deep concentration, a furrowed brow, no socializing with anybody. I was gathering thoughts for a work day. Now with him I fell into a new habit, going westward three blocks to Cooper Park, a grassy space much favored for the (illegal) leash-free cavorting of dogs. The dogcatcher doesn't patrol this park; it's a scofflaw zone by default and consensus. Here I would unclick Wiley from his leash and let him gallop off to socialize and wrestle and scratch and pee and sniff and defecate (yes, I always had a bag) as he pleased. Meanwhile I would stand jawing with the other dog people—such as Bill, the president of the local Optimists Club; or Marty, the fishing guide who summers in Alaska; or Dave, the retired neurosurgeon with the white ponytail; or Henry, the tall film student; or Barb, the nurse with the broken leg—as they too killed time while indulging their animals: Rosie and Dusty, the hyperkinetic goldens; Kvichak, the Samoyed; Merlin and Diesel, black Labs; and Frieda, the doleful Saint Bernard. These were good people and good dogs I wouldn't have known if Wiley hadn't dragged me into their company.

After a few minutes of romping and tussling, Wiley would return to my side and sit, listening patiently, as though to say: *All right, the doggy games are fine, but ultimately I prefer the company of grownups*. Then we'd go home. My wife used to tease us about these very regular morning sessions. My god, she'd say, now I'm surrounded by *two* males addicted to routine.

So we became part of the Cooper Park scene, he and I. And when someone hatched the idea of donating a water fountain to the park for both humans and (by way of a lower dish) dogs, we became part of that too. The fountain was bought with monies gathered from dog-owners, each of whom paid fifty bucks for the privilege of having one canine name engraved on a memorial brick, those bricks to form a cobbling around the fountain. One of the bricks, as eventually laid, said "WILEY." I took him to admire it when the work was done. He was blasé.

Now that he'd settled into domestic life and stopped chasing horses, there was just one problem with Wiley: the problem of time. He had passed his eleventh birthday. He acted young but wasn't. And so one day he died.

It was sudden, and therefore merciful to him and shocking to us. It happened six weeks ago as I write this, the day following our annual hockey-team party, which he had attended in full vigor with his usual keen attunement to the presence of friendly humans and loosely guarded meat. Next afternoon he went to sleep in the back-yard during a snowstorm, one of his favorite things, and didn't wake up. I found him limp, looking peaceful, going cold. Tears streaming, I wrapped him in our best blanket and wondered how on Earth to tell my wife when she returned from an evening out.

She mourned him with a depth of pain that put me in mind of Greek tragedy—maybe Hecuba over the corpse of Hector. He'd been with her, like a child, but also as protector and closest confi-dant, for ten years. I've never seen anyone love a dog more than she loved him. And me, I mourned him in sidelong, bashful ways of my own. We knew that we weren't the only people who had ever cried over a dead dog they'd considered the finest and noblest creature in the history of canines. We knew, yes; but this was our turn.

For the first few days, we hunkered. Word of Wiley's death spread to our friends and acquaintances somewhat faster than we had heart to spread it ourselves. Cards of condolence arrived. One of my hockey teammates and his wife sent flowers. In the hockey cul-ture, that's bold sensitivity. Meanwhile I avoided Cooper Park, at least during the usual hours. I had to leave town on a weekend trip to California anyway. Before departing, I made a brief visit to the park, almost furtively, and did a corny, sentimental thing—I laid a daffodil from our garden on the brick inscribed WILEY. I figured the flower itself would disappear quickly amid bustling people, bicycles, baby carriages, and dogs, but the gesture felt right.

Another week or more passed before I steeled myself to revisit the park. I was driving by, noticed a cluster of familiar figures, and on impulse I stopped. It was Dave the neurosurgeon, Bob the optimist, and Barb the nurse, along with Merlin and Rosie and Dusty and Frieda. As the three people saw me walking up, alone, their faces went long and grim. And now I have come to the point of this essay: the point when I recognized as never before that I don't live just in a house and a state and a town; that I live in a community. We saw the flower, Dave said.

Clone Your Troubles Away

ONE MORNING LAST WINTER a small item appeared in my local newspaper, announcing the birth of an extraordinary animal. A team of researchers at Texas A&M University had succeeded in cloning a whitetail deer. Never before done. The fawn, known as Dewey, was developing normally and seemed to be healthy. He had no mother, just a surrogate who had carried his fetus to term. He had no father, just a "donor" of all his chromosomes. He was the genetic duplicate of a certain trophy buck out of south Texas whose skin cells had been cultured in a laboratory. One of those cells furnished a nucleus that, transplanted and rejiggered, became the DNA core of an egg cell, which became an embryo, which became Dewey. So he was wildlife, in a sense, but in another sense elaborately synthetic. This is the sort of news, quirky but epochal, that can cause a person with a mouthful of toast to pause and marvel. What a dumb idea, I marveled.

North America contains about 20 million deer. The estimate is a rough one (give or take, say, 5 million), since no one could ever count them. Some biologists suspect that the number is higher now than it was five hundred years ago, reflecting the impacts of European settlement on the American landscape. Predators eradicated, old forests cut or thinned, more second growth, more edges and meadows—these changes are happy ones for deer. By any measure we've got plenty, and they're breeding like gerbils, poaching lettuce from suburban gardens, overflowing onto highways to become roadkill. Of the two species, mule deer and whitetail, the whitetail (*Odocoileus virginianus*) is more widely distributed and abundant—more abundant, in fact, than any other large wild mammal on the continent. Given such circumstances, it struck me as odd that someone would use postmodern laboratory wizardry to increase the total. Odder still to increase it, at some considerable cost, by just one. Cloning is expensive. Deer, I imagined, in my ignorance, are cheap.

The news item, drawn from wire services, was only a column filler that didn't offer much detail. It barely alluded to the central question: *Why* clone a deer? It mentioned that Dewey had been born back in May, seven months earlier, his existence kept quiet pending DNA tests to confirm his identity as an exact genetic copy. That settled, he could now be presented to the world. Dr. Mark Westhusin, of the College of Veterinary Medicine at Texas A&M, spoke for the team that created the fawn, explaining fondly that Dewey had been "bottle-fed and spoiled rotten his whole life." The item noted that A&M, evidently a leading institution in the field, had now cloned five species, including cattle, goats, pigs, and a cat.

One other claim in this little report (which had the flavor of a reprocessed press release) went unexamined and unexplained: "Researchers say the breakthrough could help conserve endangered deer species." Seeing that, I began planning a trip to Texas.

The notion that cloning might help conserve endangered species has been bandied about for years. Very little such bandying, though, is done by professional conservationists or conservation biologists. One lion biologist gave me a pointed response to the idea: "Bunkum." He and many others who study imperiled species and beleaguered ecosystems view cloning as irrelevant to their main concerns. Worse, it might be a costly distraction, diverting money, diverting energy, allowing the public to feel some bogus reassurance that all mistakes and choices are reversible and that any lost species can be recreated using biological engineering. The reality is that when a species becomes endangered, its troubles are generally twofold: not enough habitat and, as the population drops, not enough diversity left in its shrunken gene pool. What can cloning contribute toward easing those troubles? As for habitat, nothing. As for genetic diversity, little or nothing, except under very particular circumstances. Cloning is copying, and you don't increase diversity by making copies.

Or do you? This assumption, like the one about cheap deer, turns out to merit closer scrutiny.

The people most bullish on cloning are the cloners themselves, a correlation that's neither surprising nor insidious. They don't call themselves "cloners," by the way. Their résumés speak of expertise in reproductive physiology and "assisted reproductive technologies," a realm that stretches from human fertility medicine to livestock improvement and includes such tasks as in vitro fertilization (IVF, as it's known in the trade), artificial insemination (AI, not to be confused with artificial intelligence), sperm freezing, embryo freezing, embryo transfer, and nuclear transfer (which refers to the information-bearing nucleus of a cell, where the chromosomes reside, not the energy-bearing nucleus of an atom). There's also a process called ICSI (pronounced "icksy"), meaning intra-cytoplasmic sperm injection, helpful to elderly gentlemen whose sperm cells can no longer dart an egg with the old vigor. The collective acronym for all such assisted reproductive technologies is ART. To its practitioners, cloning is just another tool in the ART toolbox.

These ARTists are smart, committed people. Like others who feel a vocational zeal, they do what they believe in and believe in what they do. Blessed is the person so situated. But in their enthusiasm for cloning research, in their need to justify their time and expenditures to boards of directors, university deans, and the public, they send their imaginations to the distant horizon for possible uses and rationales. See what cloning could do for you, for society, for the planet? Some of the applications they propose are ingenious and compelling. Some are tenuous and wacky. Three of the more richly peculiar ones, each fraught with complexities and provocations, are cloning endangered species, cloning extinct species, and cloning pets. College Station, Texas, home of Dewey the duplicate deer, is where I picked up the sinuous trail that interconnects them.

"So this guy brought these testicles to me," says Mark Westhusin, as we sit in his office at Texas A&M's Reproductive Sciences Laboratory, on the edge of campus. The testicles in question, he explains, came from a big whitetail buck killed on a ranch in south Texas. The fellow had got hold of them from a friend and, intending to set himself up as a "scientific breeder," hoped that Westhusin could extract some live semen for artificial insemination of his does.

Westhusin, an associate professor in his mid-forties, is an amiable man with a full face and a fashionably spiky haircut. He has already explained to me about "scientific breeders," the term applied to anyone licensed by Texas for the husbandry of trophy-quality deer. Deer breeding is a serious business in Texas, where the whitetail industry accounts for $2.2 billion annually and where open hunting on public land is almost nonexistent, because public land itself is almost nonexistent. Most deer hunts here occur on private ranches behind high fences, allowing landowners to maintain—and to improve, if they wish—their deer populations as proprietary assets. Texas contains about 3.5 million whitetails, some far more valuable than others. An affluent hunter, or maybe just a passionate one, might pay $20,000 for the privilege of shooting a fine buck. A superlative buck, a giant-antlered prince of the species, can be worth $100,000 as a full-time professional sire. And the market doesn't stop at the Texas border. Westhusin has heard of a man who had a buck—it was up in Pennsylvania or someplace—for which he'd been offered a quarter million dollars. He didn't take it, because he was selling $300,000 worth of that buck's semen every year. Such an animal would be considered, in Westhusin's lingo, "clone-worthy."

Now imagine, Westhusin tells me, that they're collecting semen from that deer one day, and the deer gets stressed, and it dies. Damn. So what do you do? Well, one answer is you take cells from the dead buck and then clone yourself another animal with the same exact genotype. While you're at it, you might clone four or five.

"You're certainly not going to go out and clone any old deer just for the sake of cloning it," he says. Then again, when you're practicing—when you're developing your methods on a trial basis—you won't wait for the Secretariat of whitetails. The buck from south Texas, the one whose testicles landed in Westhusin's lab, wasn't superlative but it was good, and the experiment evolved haphazardly.

Working with his students to extract the semen, Westhusin suggested also taking a skin sample from the buck's scrotum, on the chance they might find a use for it. "We'll grow some cells," he said, "and maybe later on, if we have the time and the money, we'll do a little deer-cloning project." Eventually the effort produced a few dozen tiny embryos, which were transferred into surrogate does, resulting in three pregnancies, one of which yielded a live birth. That was Dewey, born May 23, 2003, to a surrogate mother known as Sweet Pea. The donor buck remained nameless.

The fellow who brought in the testicles remains nameless too, at least as the story is told by Mark Westhusin. "People don't want it to get out that they've got these huge, huge deer on their ranch. Because then the poaching gets so bad." Down in south Texas, people circle their land with high fences not just to keep the deer in but to keep the poachers out.

Dr. Duane C. Kraemer, a senior scientist and professor at the Texas A&M veterinary college, is also sometimes called Dewey, though not by visiting journalists or staffers reluctant to presume. He's a gentle, grandfatherly man with pale eyes and thinning hair, whose casually dignified style runs to a brown suit and a white pickup truck. His ART specialty is embryo transfer, and that's the step he oversaw on the deer-cloning project.

Kraemer was the mentor of Mark Westhusin, who did his doctorate at A&M and then worked for several years in the private sector before returning as faculty. The relationship between academic reproductive physiologists and the livestock business tends to be close, even overlapping, because this is a practical science. There's money in assisting the reproduction of elite bulls, cows, and horses, and that money helps fund research. Kraemer himself, raised on a dairy farm in Wisconsin, has been at A&M for much of the past fifty years, during which time he took four degrees, including a Ph.D. in reproductive physiology and a D.V.M., and performed the first commercial embryo transfer in cattle.

Working on yellow baboons, a more speculative project with implications for human medicine, he did the first successful embryo transfer in a primate. He also did the first embryo transfer in a dog and the first in a cat. Embryo transfer isn't synonymous with cloning—the embryo being transferred needn't be a clone— but it's a necessary stage in the overall cloning process. Within that specialty, and beyond it, Kraemer has been a pioneer. In the late 1970s, he and colleagues engineered the birth of an addax, a rare African antelope, using artificial insemination with sperm that had been frozen. People at the time asked: Why work with addax? The species, *Addax nasomaculatus*, didn't seem endangered. Now it's extinct everywhere except for a few patches of desert in the southern Sahara. After five decades of quietly working the boundary zone between veterinary medicine and reproductive science, Kraemer is one of the patriarchs of the ART field. Dewey the deer was named in his honor.

For Kraemer, the impetus to work with wildlife came partly from his students, some of whom asked him to teach them skills that might be applied to endangered species. Semen freezing and artificial insemination were proven techniques twenty-five years ago. Embryo transfer and in vitro fertilization showed great promise. Cloning—that was a dream. Kraemer had some small grants to support the student training, but after graduation his young people faced poor odds of landing a job in wildlife or zoo work. "We had told them right up front," he says, " 'You better have another way of making a living, and you may have to do this on the side.' " Mark Westhusin, for one, took a job doing research for Granada Bio-Sciences, part of a large cattle company.

Kraemer meanwhile established an effort he called Project Noah's Ark, aimed at putting students and faculty into the field with a mobile laboratory. The lab, in a 28-foot trailer, was equipped for collecting ova, semen, and tissue samples from threatened populations of wild animals in remote settings, such as the desert bighorn sheep in west Texas. The project's three purposes were to train students, to research techniques, and to preserve frozen tissue samples for possible cloning. At present the trailer contains a surgical cradle capable of holding an anesthetized bighorn, a portable autoclave (for sterilizing instruments), a laparascope with fiber optics (for extracting ova from ovaries), three 50-amp generators, and an earnest sign: "Ask not only what Nature can do for you, but also what you can do for Nature.—D. C. Kraemer." Asking what he could do for Nature by way of assisted reproductive technologies didn't bring Kraemer much financial support. The training has gone forward, but the ark itself is in dry dock.

Animal cloning began, back in 1951, with frogs. Robert Briggs and Thomas J. King were embryologists based at a cancer research institute in Philadelphia, with a medical interest in understanding how genes are turned on and off during embryo development. Briggs, the senior man, figured that a cloning experiment might bring some insight. What he envisioned was transferring the nucleus of a frog cell, taken from an embryo, into an enucleated frog egg—that is, one from which the original nucleus had been scooped out like the pit from an olive. King, hired for his technical skills, would do the micromanipulation, using delicate scissors and tiny glass needles and pipettes. The transferred nucleus would contain a complete set of chromosomes, carrying all the nuclear DNA required for guiding the development of an individual frog. If things went as hoped, the reconfigured egg would divide into two new cells, divide again, and continue dividing through the full course of embryonic growth to yield a living tadpole. From 197 nuclear-transfer attempts, Briggs and King got 35 promising embryos, of which 27 survived to the tadpole stage. Although the success rate was low, barely one in eight, their experiment represented a large triumph. They had proved the principle that an animal could be cloned from a single cell.

Two questions followed. First, could it be done with mammals? Second, could it be done not just from an *embryo* cell, as DNA donor, but from a *mature* cell snipped off an adult? The second question is weighty, because cloning from embryo cells is, except under special conditions, cloning blind. If you don't know the adult character of an individual animal—is it healthy, is it beautiful, is it swift, is it meaty, does it have a huge rack of antlers?—why take pains to duplicate it?

For decades both questions remained in doubt. Nobody succeeded in producing a documented, credible instance of mammal cloning. One researcher claimed to have cloned mice, but his work fell under suspicion, and it could never be verified or repeated. In 1984 two developmental biologists went so far as to state, in the journal *Science*, that their own unsuccessful efforts with mice, as well as other evidence, "suggest that the cloning of mammals by simple nuclear transfer is biologically impossible."

Well, no, it wasn't—as proved that very year by a brilliant Danish veterinarian named Steen Willadsen. Unlike the developmental biologists who experimented with frogs or laboratory mice, but like Kraemer and Westhusin, Willadsen focused on farm animals. Working for the British Agricultural Research Council at a laboratory in Cambridge, he achieved the first verified cloning of a mammal. He did it—a dozen years before the famously cloned bovid, Dolly—with sheep. He took his donor cells from early sheep embryos, which had not yet begun to differentiate into the variously specialized cells (known as somatic cells) that would eventually form body parts. Such undifferentiated cells, it seemed, were crucial. Transferring one nucleus at a time into one enucleated ovum, fusing each pair by means of a gentle electric shock, following that with a few other crafty moves, Willadsen got enough viable embryos to generate three pregnancies, one of which yielded a living lamb. The following year, afloat on his reputation as a cloner, he left Cambridge for Texas, hired away by the same cattle company, Granada, that soon afterward would also hire Mark Westhusin.

"And so," Duane Kraemer says, "Dr. Willadsen then came and taught us how to do cloning."

But Willadsen couldn't teach them to clone an animal from a skin sample sliced off a buck's scrotum, because he hadn't solved the special problems of cloning from somatic cells. Between early embryo cells (which all look alike) and somatic cells (specialized as skin, bone, muscle, nerve, or any sort of internal organ) lurks a deep mystery: the mystery of development and differentiation from a single endowment of DNA. Each cell in a given animal carries a complete copy of the same chromosomal DNA, the same genetic instructions; yet cells respond differently during development, fulfilling different portions of the overall construction plan, assuming different shapes and roles within the body. How does that happen? Why? What tells this cell to become skin, that cell to become bone, another to become liver tissue? What signals them to implement part of the genetic instructions they carry and to ignore all the rest? Big questions. Cloning researchers, if they were ever to produce an animal cloned from an adult, didn't necessarily need to answer those questions, but they needed to circumvent them. They needed somehow to erase the differentiation of the donor DNA and to conjure it into operating as though its role within a living creature had just begun anew.

That's what Ian Wilmut, Keith Campbell, and their colleagues in Scotland managed in 1996, using some further touches of biochemical trickery. The result was Dolly, her existence revealed in *Nature* the following year. Dolly's donor cell came from the udder of a six-year-old Finn-Dorset ewe. Her birth was significant because it meant that cloners could now shop before they bought.

Lou Hawthorne, a cagey businessman with a trim beard, a weakness for droll language, and a soft heart for animals, tells me how the notion of dog cloning arrived at Texas A&M. Hawthorne is the CEO of a California-based company called Genetic Savings & Clone, which offers the services of "gene banking and cloning of exceptional pets." Another man, Hawthorne's chief financial backer, who prefers to avoid media attention, set the process in motion with a personal whim. "It was just one morning, he was reading the paper," says Hawthorne. "Dolly had been cloned. There was an article about it, and he said: 'I think I'd like to clone Missy. I can afford it.'"

The "he" refers to John Sperling, founder of the Apollo Group, a $2 billion empire that encompasses, among other things, the University of Phoenix, a lucrative enterprise in higher education for working adults. Missy was ten years old at the time of Sperling's brainstorm, a dog of unknown lineage but winning charms, adopted from a pound. Asked by Sperling to make inquiries, Lou Hawthorne solicited proposals from a dozen laboratories; the best came from Texas A&M.

Westhusin remembers telling Hawthorne that they could give it a try but that trying might cost a million dollars a year, take five years, and still be uncertain of success. Okay, said Hawthorne. John Sperling, as he himself had declared, could afford it. So the R&D effort toward producing a duplicate Missy—or maybe a multiplicity of copies—began at College Station in 1998. Hawthorne, a word man among scientists, named it the Missyplicity Project.

At the start it was a joint venture between Texas A&M and an earlier company led by Hawthorne, the Bio-Arts and Research Corporation. Missy contributed a patch of skin cells, which were multiplied by culturing in vitro and then frozen for future use. Westhusin's team gathered a pool of female dogs to serve as egg donors. The eggs, harvested surgically from the oviducts whenever a dog showed signs of ovulation, were emptied of their nuclei using micromanipulation tools (tiny pipettes guided by low-gear control arms within the field of a binocular scope) and then refitted with Missy's DNA by nuclear transfer. These refitted cells were nurtured in the laboratory until some of them showed good embryonic development. Promising embryos were then implanted surgically in ready (that is, estrous) surrogate mothers. Among the factors that make dog cloning difficult is that female canines come intro estrus irregularly. Unless you're keeping a riotous kennel, you may not have a bitch in heat when you need her. Westhusin and his Missyplicity partners struggled against that limitation and others for almost five years.

Missy herself died in 2002, still unitary, uncloned. But of course it isn't too late. Her genome is on ice.

Meanwhile, two interesting new entities were born in College Station. One was a cloned cat, the world's first, a little domestic shorthair kitten given the name CC, standing for "copycat." The other was Hawthorne's present company, Genetic Savings & Clone, a for-profit operation devoted to the gene-banking of pets (in the form of frozen cells) toward the possibility of their eventual cloning. CC, created with nuclear DNA from a calico donor named Rainbow, was delivered by cesarean section just before Christmas of 2001. GSC came into being in response to popular demand.

Alerted to the Missyplicity Project by press reports, dog and cat owners had begun contacting the A&M lab. Some were grieving over recently deceased pets; some were concerned in advance over old animals or sick ones. "We're supposed to be focusing on research here," Westhusin recalls thinking, "and we don't have time to take fifteen phone calls a day and talk to these people about their pets." The callers tended to be emotional, poorly informed, and hopeful. *I buried him three days ago. Do you think there's any chance if I go dig him up that you could get cells off him?* "Um, I doubt it," Westhusin would say. *Well, the temperature up here is cold. It's Minnesota. . . .* Westhusin laughs pityingly, and so do I. "You want to be nice," he says, "so you sometimes spend thirty minutes talking on the phone." With the founding of Genetic Savings & Clone, all that grief counseling could be outsourced. Dr. Charles R. Long, another reproductive physiologist and an old friend of Westhusin's, was hired to get the company launched. As general manager, he recruited technical staff and established a lab to work in partnership with the A&M people. Occasionally he found himself playing psychologist to prospective customers, as Westhusin had done. "The people, the overly emotional ones— many times I would quite frankly try to convince them not to make this decision," Long says. Why? Because they were doing it for the wrong reason. "They were doing it to try to get their special animal back, and you can't get your special animal back. There's no such thing as resurrection. At least not in pets." What you get is just a genetic copy, a new animal with the old DNA, "and it's really important for people to understand that." Chuck Long is a bright, unpretentious man with a small neat mustache, the neck of a linebacker, and huge hands. He once loved a golden retriever named Tex, but he wouldn't have cloned the animal. A loving relationship is about discovery. He'd rather discover a new friend than try to relive life with Tex Two.

Where do people get their misguided ideas about cloning? I ask.

"Hollywood," says Long.

Half ignoring his answer, I press: Do they get them from scientists who oversell the technique or from the media?

"From Hollywood, I think," he repeats. "You know, crazy movies like Arnold Schwarzenegger's *The 6th Day*."

"Was he a clone in that?"

"Yeah, they cloned him."

"I haven't seen that one."

"You've got to see *The 6th Day*. It really stinks."

After a couple years at Genetic Savings & Clone, Chuck Long parted ways with Lou Hawthorne, and he now works more comfortably for a Texas company, Global Genetics and Biologicals, involved in the production and international export of elite livestock. GSC itself has severed its relationship with Texas A&M and relocated its headquarters to Sausalito, California, with offices overlooking a kayak beach.

Genetic Savings & Clone isn't the only company that sells a gene-banking service for pets; you might also turn to Lazaron BioTechnologies or PerPETuate, Inc. But GSC alone offers the full deal: delivery of clonal duplicates in the near future. The initial cost of putting your pet's genes into the gene bank is $895. Annual storage runs $100. Dogs, with their unique physiological complications (such as opacity of the eggs, making them harder to enucleate), are still problematic. But commercial cat cloning got under way in 2004, with five clients committed, and if all goes well, their cats will be delivered very soon. "In pet cloning," says Hawthorne, "people have an animal that they perceive is extraordinary. In some cases, it's just a perception." In other cases, the extraordinariness is more objective. "You can have an extraordinary mutt," he says. "You can have an animal that has extraordinary intelligence. Extraordinary good looks." Insofar as those traits are genetic, they can be reproduced by cloning, maybe. The delivery price of a healthy young feline, custom-created from DNA of proven appeal, guaranteed to resemble your old feline closely, is $50,000. If you think this might meet your emotional expectations, act now.

On the other hand, $50,000 buys a lot of pretty good cats.

Cloning endangered species is a different matter. For starters, who pays? Why does anyone finance this technical approach, seemingly so marginal, rather than putting money toward basic necessities such as habitat protection? And how can cloning possibly freshen a gene pool that has been reduced to a stagnant puddle? I carry these questions to Dr. Betsy Dresser, director of the Audubon Center for Research of Endangered Species, near New Orleans. ACRES is part of the Audubon Nature Institute, a nonprofit group of museums, parks, and other facilities (with no connection to the National Audubon Society). Dresser, who ran a similar research center at the Cincinnati Zoo before coming to New Orleans, has long been a leader in applying captive-breeding efforts and assisted reproductive technologies to endangered species.

She's a brisk, congenial woman, but not easy to get to. ACRES is tucked away in a sunny new building surrounded by bottomland forest at the end of a country road outside the city, on the west bank of the Mississippi River, beneath a towering levee. The land, 1,200 acres of what once was sugar plantation, is protected by a fence and a guard house with an electric gate. A sign says FREE-PORT MCMORAN AUDUBON SPECIES SURVIVAL CENTER, recognizing the sponsorship of a mining company in establishing this compound. ACRES itself was created with a $15 million appropriation from the U.S. Fish and Wildlife Service. It resembles the visitor's center of a well-funded state park, but more private. On a morning in April, the air is redolent with honeysuckle. I arrive in time to watch surgery on a domestic cat.

Dr. C. Earle Pope, in a blue smock and mask, is harvesting ova. Several other figures, also in blue, assist him around the operating table. The cat has already been anesthetized and opened, its ovaries exposed. Pope wields a fine forceps in one hand, a hollow steel needle in the other, his head raised to view the target area as magnified on a video monitor. He works with easy skill derived from years of experience. The hollow needle is backed by a suction device that feeds into a glass vial on a table nearby. With the forceps, Pope gingerly lifts one ovary so that its follicles (the small, bulbous ovarian sacs) protrude like grapes on a bunch. With the needle, he punctures a follicle and sucks out the egg, then moves to another. The ovary bleeds slightly. Poke, suck, poke, suck, the eggs are whisked away. They accumulate in the vial. When Pope has emptied the follicles of both ovaries, an assistant collects two orange-caped vials and passes them through a window from the operating room to an adjacent lab.

In the lab, which is darkened and barely larger than a closet, a technician moves the eggs from a rinsing solution onto a petri dish. She lifts them one by one, using an aspirator pipette—that is, with suction applied by her own gentle breath. Her eyes are pressed to a scope. The eggs, surrounded by cloudy globs of ovarian material (called cumulus cells) and air bubbles, are tiny, but they are conspicuous enough to her. You can tell the maturity of the ova, she says, by the layers of cumulus cells attached. "These are very good-looking." The yield today is twenty-four eggs, about average from a domestic cat. Down the hall, she places the petri dish in an incubator. This afternoon one of Pope's colleagues, Dr. Martha C. Gomez, will enucleate these eggs and endow each with nuclear DNA transferred from an African wildcat.

It won't be the first time such a mix is performed. The Africa wildcat, *Felis silvestris lybica*, is a tawny little felid native to Africa and the Middle East, related to the domestic cat, *Felis silvestris catus,* closely enough to have figured in earlier experiments involving the two subspecies. Using domestic cat eggs, Gomez, Pope, and their team produced three Africa wildcat clones in 2003, the eldest born on August 6 and named Ditteaux. (That's *ditto* with Cajun spicing.) The animal from which he and his . . . his what? not siblings, not twins—his two extremely close relatives were cloned, known as Jazz, was itself a product of combined ARTs: the world's first frozen-embryo, thawed-embryo, embryo-transferred wildcat born to a domestic cat. Gomez, Dresser, Pope, and several colleagues coauthored a journal paper on this work, in which they note that the African wildcat "is one of the smallest wild cats, whose future is threatened by hybridization with domestic cats." A person might ask: If hybridization of a wild subspecies with a domestic subspecies is the threat, in what sense is mixing nuclei from one subspecies with eggs from the other subspecies a solution?

Another skeptical question, which I put to Betsy Dresser, is whether this fancy stuff can somehow mitigate the problem of low genetic diversity in sorely endangered species. If it can't, what's the point? "Well, indeed it can," Dresser says. "What we're trying to do is use cloning to bring in the genetic material from animals that are not reproducing." Among any population, she says, there are always infertile individuals, marginalized individuals, elderly or unlucky individuals, who fail to breed and so contribute no genes to the next generation. In a large population (though Dresser doesn't mention this point), their exclusion represents Darwinian selection, which drives evolution. But in a very small population (she notes rightly), their participation could be crucial. "If you can use the genetic material from those individuals, it helps widen the genetic pool a bit."

Imagine you've got a captive population of just five black-footed ferrets, with no others surviving on the planet. Four of your ferrets are males and the fifth is a postreproductive female. One young male chokes to death while eating a prairie dog with reckless gusto. What do you do? Of course you grab the old female and the dead male, take tissue samples, and clone them. But wait—in this scenario of five, there are no viable black-footed ferret ova to receive the clonal DNA. So you use the next best thing: enucleated eggs from a mink. Then you breed your cloned female with one of the males, breed any daughters she produces with other males, get the cloned male's genes into the reproductive jumble too, and thereby postpone (maybe indefinitely) the doom of your miserable little population. Whether your ferrets ever go back into the wild is another question. Do you dare send them? Do you keep breeding and cloning until you've got a few dozen, a few hundred? All this would be expensive at best and, if you hadn't meanwhile solved the root causes of endangerment (such as insufficient habitat, government-sponsored poisoning of prairie dogs, poaching, or exotic species inflicting too much predation or competition), ultimately futile. No clones of an endangered species, and no descendants of clones, have ever yet been released to the wild.*

What about the money issue? I ask Dresser. Are resources being diverted that might otherwise pay for habitat preservation? Her answer is candid: "The money that comes to this kind of research

---

* It bears noting, though, that other ART methods have yielded some returns to the wild, such as the hundred Mississippi sandhill cranes, additions to an endangered subspecies, that were bred at ACRES using artificial insemination and released on a refuge near Pascagoula.

is primarily from people that are not going to support habitat." She's a skilled fundraiser as well as a respected scientist; she has been through this in Cincinnati, now New Orleans, and she knows her constituency. Sponsoring the research arm of a fine metropolitan zoo is a bit like sponsoring the symphony, the conservatory, the opera. These people "don't want to give their money to Africa, or to Asia, or somewhere. They don't want their money in political environments where they're never going to see their name on a plaque." At the various branches of the Audubon Nature Institute, including ACRES, there are more than a few grateful plaques.

Back on the city side of the river, I visit the Audubon Zoo on Magazine Street for a glimpse of Ditteaux the cloned wildcat, temporarily on display there. For this interlude of public exposure, he lives in a glass-fronted cage furnished with small boulders, trees, and a six-foot square of scenery meant to approximate northern Africa. He's a handsome animal, lanky and lithe, nervous, his brownish gray fur marked with pale stripes. As I watch, his pale green eyes come alert to something—the sight of a squirrel outside the building, visible through an opposite window.

Groups of schoolchildren pass Ditteaux's cage. A well-fed boy in an orange T-shirt reads the sign and then asks, "It's a clone?" Yes. With some vehemence, he says, "Okay, that's *freaky*."

Whatever the downside of investing money and time in such an approach to endangered species, at least one private firm has also done it: Advanced Cell Technology, of Worcester, Massachusetts. Founded originally as a subsidiary of a poultry genetics business, ACT now concentrates mainly on human and medical issues. The company's work with wildlife is an adventuresome sideline, bearing no such commercial promise as cloning whitetail deer for the trophy market but offering the possibility of a public good, roughly equivalent to pro bono work by a law firm. It also offers, when successful, good publicity.

In early 2001, ACT announced that "the first cloned endangered animal," an eighty-pound male gaur, had been born to a surrogate mother. The gaur is a species of wild cattle, *Bos gaurus*, native to southeastern Asia from Thailand to Nepal. Calling it an "endangered animal" was mildly misleading; the international body that keeps track of such things classifies the gaur as "vulnerable," not actually "endangered," with somewhere between 13,000 and 30,000 individuals in the wild. But the population is declining, and the trend isn't likely to reverse. Vulnerable or endangered, the species deserves attention.

Two technical points made ACT's gaur work especially notable. First, the nuclear DNA came from gaur cells derived from a tissue sample that had sat frozen for eight years in a gene bank at the San Diego Zoo. Second, the enucleated egg cell into which the gaur DNA had been transferred came from a domestic cow. So this too was a case of cross-species cloning—in fact, it was the first recorded case, precursor to the African wildcat project in New Orleans. Arguably, the technique could be valuable in situations when egg cells of an endangered species are unavailable—when there are no surviving females, say, or so few that you wouldn't dare cut one open to harvest her eggs.

What made the case less encouraging was that the baby gaur, named Noah, died of dysentery within two days. Its death fell hard on Robert P. Lanza, a vice president of ACT, who had led the cloning effort.

At that time, Lanza had nearly sealed an agreement with Spanish officials toward cloning an extinct Spanish subspecies of mountain goat, the bucardo. The bucardo (*Capra pyrenaica pyrenaica*) had languished at desperately low population levels throughout the twentieth century, probably because of competition with livestock, diseases caught from livestock, poaching, and other travails. The last one died in 2000, clunked by a falling tree, but provident biologists had arranged to freeze some of its tissue for posterity. Lanza hoped to clone the bucardo back into existence, using the frozen sample for nuclear DNA, a domestic nanny goat as egg donor, and a nanny again as surrogate to carry the fetus. That plan collapsed with the death of Noah the gaur. Two years later ACT's cloning team tried again, this time achieving the birth of two cloned calves from another species of wild Asian cattle, the banteng, *Bos javanicus*. The banteng is unambiguously endangered, with no more than 8,000 individuals in the wild. The nuclear DNA came from another frozen sample that had been stored, for twenty-five years, at the San Diego Zoo.

The gene bank in San Diego, loosely known as the Frozen Zoo, was established three decades ago by a pathologist named Kirk Benirschke, who was soon joined by a young geneticist, Oliver A. Ryder. Benirschke and Ryder foresaw that these cell samples might be useful in genetic studies of relatedness among wild species. They didn't foresee that the frozen cells might be cloned back to life. The collection now represents about 7,000 individual animals of 450 different species; about half of those samples came from creatures resident at the San Diego Zoo, the rest from other zoos and captive facilities, or from the wild. Ryder is still there, the man to see if you want a morsel of rare or endangered DNA for some legitimate purpose. Cloners across the country, from College Station to Worcester and beyond, point to San Diego's Frozen Zoo as a prescient enterprise that should be emulated widely, preserving as much genetic diversity as possible from endangered species before their populations decline too far. Ryder, for his part, supports the idea of cloning when it might return a valuable genotype to a breeding population. The original banteng whose frozen cells went to ACT, for instance, died in 1980 without offspring, having made no genetic contribution to the captive banteng population. One of the two clones produced from those cells was healthy, and that animal has since been returned to San Diego; its lost genes may eventually be bred back into the zoo population of banteng, possibly adding some much-needed diversity.

But gene banking is no panacea. Ryder himself says: "I think it's gonna be a somber day when we realize that the only thing left of a species is something we've got in the Frozen Zoo."

Among extinct species and subspecies, the bucardo goat represents a good prospect for cloning, because the extinction is so recent and the cell sample was properly preserved. Less propitious circumstances, though, don't prevent people from trying to resurrect a lost beast.

Scientists at Kinki University in Japan have begun work toward cloning a woolly mammoth, using tissue samples from a 20,000-

year-old carcass recently excavated from frozen Siberian tundra. Elephants, the mammoth's closest living relatives, will serve as egg donors and surrogate mothers, if the project ever gets that far. Cloning researchers at the Australian Museum in Sydney hope to recreate the thylacine, a carnivorous marsupial loosely known as the Tasmanian tiger, last seen alive in 1936. For that effort, the starting point is a thylacine pup stored in alcohol since 1866. Alcohol is a gentler preservative than formaldehyde, and the Australians have managed to extract some DNA fragments in fairly good condition—but no complete DNA strands, let alone any viable thylacine cells with nuclei that could be transferred intact. The optimistic Aussies aim to reassemble their squibs and scraps into a full set of thylacine DNA, perhaps patching the gaps with genetic material from other marsupials. Plausible? Not very, according to Ryder. "What's the chance that you could shred the phone book," he asks, "and then drop it out of a window and have it come back together?" Once they have reassembled their phone book, if they do, the Australians will create artificial chromosomes for insertion into an egg from some related species, such as the Tasmanian devil. Meanwhile, in Hyderabad, India, a team led by Dr. Lalji Singh proposes to clone an Asiatic cheetah, a subspecies extinct in India for the past fifty years. They want to use nuclear DNA from a cheetah loaned by Iran, though Iran itself has only a few dozen cheetahs in the wild, and none of those has been promised so far. If the Indians do get their chance to proceed, the eggs and the surrogate wombs will be furnished by leopards.

Each of these projects, variously dreamy or doable, represents an effort at cross-species cloning, like the banteng-and-cow work by ACT. This sort of trick raises further issues. What are the physiological consequences of mixing nuclear DNA from a cheetah with mitochondrial DNA (which comes along with the enucleated egg and helps regulate the cell's biochemistry) from a leopard? What are the ecological implications of mixing mammoths with elephants in a world where the mammoths' ecosystem no longer exists? What's the merit or demerit of blurring lines between species (cheetah and leopard, thylacine and devil) by means of laboratory gimmickry, in order to "preserve" a vanishing subspecies or "restore" an extinct species in the wild?

Lines, their integrity or transgression, are exactly what's at issue: the line between one species and another that defines biological diversity, the line between one animal and another that constitutes individuality, the line between living and dead that gives meaning—as well as poignant temporal limit—to life. And yet those lines aren't always easy to draw, let alone to enforce or respect. Even species, even in the wild, sometimes blur into one another: wolves breeding with coyotes, blue-winged warblers with gold-winged warblers, barn swallows with house martins, mule deer with whitetails. True, these natural mongrelizations represent exceptions to the rule of how species are generally demarcated. But they complicate any efforts to think clearly about drawing other lines, such as the line between *Felis silvestris lybica* and *Felis silvestris catus*, the line between embryo transfer and nuclear transfer, the line between genetically modified organisms and heirloom tomatoes (which have themselves been genetically modified by generations of careful horticulture), the line between extinct and merely frozen, the line between what we can do and what we should do, the line between nature and ART.

Recognizing such complications is not necessarily the same as surrendering to a paralyzing relativism. Lines that suggest boundaries of ethical behavior, of judicious balance between opposing concerns, and of precious entities deserving preservation are important even when they reveal themselves, at close inspection, to be smeary zones of gradated gray. The mapping of such boundaries can't be done by science, which is capable of measuring shades of gray but not choosing among them. That leaves religion, philosophy, social consensus, and common sense. Which of those do we rely on for decisions about assisted reproductive technologies, such as cloning, when the species being assisted is not the banteng or the whitetail deer but *Homo sapiens*?

Consider the prospect of germline genetic engineering—that is, fiddling with genes in human embryo cells before those cells are grown into human fetuses. Germline engineering is not yet available as a consumer option, for medical purposes or any others, but soon it may be. Select genes would be added to, subtracted from, or modified in an embryo cell, after which the cell would be cloned into a customized human child. This process would permit the correction of genetic weaknesses—bad eyesight, for instance, or sickle-cell anemia—in advance of birth. When that starts happening, as Bill McKibben has warned in his book *Enough: Staying Human in an Engineered Age*, "the line between fixing problems and 'enhancing' offspring" will disappear, at least for any parents who want their kids to be as bright, robust, good-looking, and competitive as humanly (that is, technologically) possible. If you can repair your future child's myopia with preemptive genetic tinkering, you might also want to increase her IQ by a few dozen points. Will it lead to a world as utopian as Lake Woebegon, where all the children are above average? Of course not. It will just add genetic manipulation of embryos and child cloning to the means by which affluent, fussy people try to distance themselves from bad luck, disappointment, menial work, death, and poor people.

McKibben, his ardent humaneness informed by a lot of careful research and thinking, proposes that we should recoil from such possibilities and declare "Enough!" He suggests that somewhere amid the dizzying possibilities of ART as applied to humans, beyond fertility medicine but short of germline genetic engineering, we might locate "the enough line"—that is, the threshold of ugly and corruptive weirdness across which a wholesome person and a wise society do not go.

As much as I want to agree with him, my own survey of animal cloning forces me to conclude that his "enough" line, like any I might try to draw myself, is as subjective as it is sensible. There is in fact no line. There is only a spectrum, a set of choices among shades of gray. Of course, that's not to say some choices aren't nuttier than others.

Cloning adult humans, for instance. Any thorough discussion of assisted reproductive technologies comes eventually to this topic, which the animal-cloning scientists detest and dismiss but which other people consider central. The animal guys are right—it's not central—but like a parrot in a cage of canaries, it's too big and noisy to ignore. What if John Sperling or some other loopy billionaire decides one morning to commission not the cloning of his lovable mutt but the cloning of himself? If that decision hasn't already been made, quietly in a penthouse somewhere, it probably soon will be; and whatever unique technical difficulties or scientific scruples have so far prevented the consummation of such a desire will soon be overcome. Some people view the prospect of human cloning with great alarm. Bill Clinton labeled it "morally reprehensible." His presidential ethics commission recommended federal laws to prohibit human cloning. Finding myself less certain than Clinton or those advisers about the moral or legal verities against which human cloning should be measured, I'd simply call it perniciously stupid. Then again, many things people do nowadays are, in my opinion, perniciously stupid. Not all of them are illegal, and so, I suppose, human cloning needn't be either.

Down in College Station, I'm reminded of all this during my chat with Duane Kraemer, when we bounce from the subject of endangered species back to companion animals. Isn't there something misguided, I ask Kraemer, about cloning your pet? Doesn't it reflect an inclination to deny mortality?

Deny mortality? "We do that every day!" he says brightly. "We get up and brush our teeth. Why do we do that? Because we want to live as long as we can. So denial of mortality is, yeah, it's in our being. And it's not only natural. It's necessary."

Two other voices of wisdom echo through my head, addressing aspects of the question *why*. One of these voices belonged to J. Robert Oppenheimer, the physicist and founding director of the Los Alamos nuclear weapons laboratory. Trust me on this seeming digression. Having helped build the first atomic bomb, Oppenheimer resisted the notion that America should rush ahead to build a thermonuclear superbomb. It was fission versus fusion, uranium versus hydrogen, kilotons versus megatons, and the global political context of 1943 versus the context of 1951. His resistance was swept aside by a clever design principle concocted by two other physicists, one of whom was Edward Teller. Asked later by an inquisitorial panel about how the H-bomb decision was made, Oppenheimer declined to speak about technical details. "However," he said mordantly, "it is my judgment in these things that when you see something that is technically sweet, you go ahead and do it and you argue about what to do about it only after you have had your technical success." This scary truth, which might be thought of as Oppenheimer's Axiom, explains many controversial gambits in whizbang scientific engineering. Why do some scientists crave to clone animals? Not just because they can but because they can do so, with an elegant medley of ingenious laboratory moves, in a way that is technically sweet. And therefore irresistible.

The other voice comes from Louis Armstrong, as recorded in 1931:

> *Oh, when skies are cloudy and gray,*
> *They're only gray for a day,*
> *Bay-bay-bay-bee . . .*
> *So wrap your troubles in dreams,*
> *And dream your troubles away.*

Duane Kraemer is right in noting that this problem-solving approach isn't unique to assisted reproductive technologists.

On the morning after our conversation, Dr. Kraemer welcomes me to his home, in a neighborhood just north of the A&M campus, to meet the famous cloned house cat, CC. As we enter, she crosses a living room of draped-over furniture and leaps onto a carpeted cat perch, presenting herself for Kraemer's gentle petting. She's no longer a kitten. She arches her back to my touch, then carefully sniffs my hand. Her fur is soft and clean. She looks like any normal cat. The most striking aspect of her appearance, which I wouldn't notice if I hadn't read some background, is that she's a tiger-tabby shorthair, mottled black-and-gray with a white chest and legs. It's striking because she was cloned from a calico.

That is, CC's color pattern differs utterly from that of Rainbow, her DNA donor. The cause of this difference is complicated (involving random inactivation of one of her two X chromosomes, which in a female such as CC are redundant, though each may carry a distinct gene for color). But those complications can be reduced to a single, simple word: random. The application of one color program and the inactivation of the other, in such circumstances, is determined by chance. And by chance CC's coloring is unlike

Rainbow's. Cloning isn't resurrection, as the man said. It isn't even, quite, duplication.

On CC's right cheek, otherwise white, I notice a small patch of tan fur, like a birthmark. Yes, says Kraemer, that wasn't present in Rainbow either. The genotype may be identical in a clone, but it gets expressed differently. Maybe one day when she was a fetus, inside the surrogate mother, CC rubbed her little face against the wall of the womb. A smudge. Things happen.

AUTHOR'S NOTE

First publication of each of the pieces was as follows: "Sympathy for the Devil," *Outside* (June-July 1981); "Has Success Spoiled the Crow?," *Outside* (October 1983); "The Widow Knows," *Outside* (April 1982); "The Troubled Gaze of the Octopus," *Outside* (July 1984); "Avatars of the Soul in Malaya," *Outside* (March 1984); "Rumors of a Snake," *Outside* (April 1984); "Wool of Bat," *Outside* (August-September 1982); "The Excavation of Jack Horner," *Esquire* (December 1984); "The Lives of Eugène Marais," *Outside* (October 1981); "The Man with the Metal Nose," *Outside* (June 1983); "Animal Rights and Beyond," *Outside* (June 1984); "Alias Benowitz Shoe Repair," *Outside* (December 1983); "The Tree People," *Outside* (January-February 1984); "Love's Martyrs," *Outside* (September 1983); "A Deathly Chill," *Outside* (December-January 1983); "Is Sex Necessary?" *Outside* (October 1982); "Desert Sanitaire," *Outside* (February-March 1983); "Jeremy Bentham, the *Pietà*, and a Precious Few Grayling," *Audubon* (May 1982); "Yin and Yang in the Tularosa Basin," *Audubon* (January 1985); "Planet of Weeds," *Harper's* (October 1998); "The River Jumps Over the Mountain," *National Geographic Adventure* (February 2002); "The Post-Communist Wolf," *Outside* (December 2000); The Megatransect series, I. "Into the Forest," *National Geographic* (October 2000), II. "The Green Abyss," *National Geographic* (March 2001), III. "End of the Line," *National Geographic* (August 2001); "A Passion for Order," *National Geographic* (June 2007); "Citizen Wiley," *Smithsonian* (October 2006); "Clone Your Troubles Away," *Harper's* (February 2005).

Many people help a magazine writer along his way. The pieces gathered here span more than twenty-five years of effort, and in that time I've accumulated debts of gratitude to more patient scientists, unsuspecting subjects, editors, friends, loved ones, colleagues, consulting experts, and casual informants than I can list here or remember. I reaffirm my thanks to all those mentioned in the first edition of this book, and I add thanks to the many more who have been helpful and generous to my later work—in particular these trusted and trusting editors, with whom I've worked in recent years: Luke Mitchell, Colin Harrison, and Lewis Lapham at *Harper's*; Oliver Payne, Kathy Moran, Bill Allen, and Chris Johns at *National Geographic*; Rebecca Maksel at *Smithsonian*; Hal Espen at *Outside*; Steve Byers and John Rasmus at *National Geographic Adventure*. Maria Guarnaschelli at W. W. Norton has once again been my vital editorial partner in shaping these pieces into a book. My deep gratitude to Renée Wayne Golden won't end with her retirement. And it was Betsy Gaines who brought Wiley, as well as so much else, into my life.

PARTIAL SOURCES

As the years passed, my style of research evolved, and the progressively greater abundance of sources consulted for each piece, as recorded here, reflects that. One thing hasn't changed much: I've always favored careful reading of the scientific literature—the journal papers scientists write for one another—over the option of simply telephoning experts and asking them to explain or comment. When I do call scientists, it's usually to ask them: May I come to see you, watch you, talk with you within your working context? Such contacts are in many cases recounted in the texts of the pieces reprinted in this book. I've found that most scientists, though they have good reasons to be wary of the popular press (reasons such as chronic inaccuracy, slapdash research, oversimplification, failure to double-check facts), are quite helpful and forthcoming if you've done your homework (again, that means reading the literature) before you begin making demands on their time and patience.

In a very few cases, among the citations below I have shown the original date of publication (in parentheses) as well as the date of the edition I used. Those cases are books, such as Playfair's popularization of James Hutton, for which the original date reflects a particular historical context within which the work should be understood.

Sympathy for the Devil

Bates, Marston. 1949. *The Natural History of Mosquitoes*. New York: Macmillan.

Gillett, J. D. 1972. *The Mosquito: Its Life, Activities, and Impact on Human Affairs*. New York: Doubleday.

Harrison, Gordon. 1978. *Mosquitoes, Malaria and Man: A History of the Hostilities Since 1880*. New York: Dutton.

Horsfall, William R. 1972. *Mosquitoes: Their Bionomics and Relation to Disease*. New York: Hafner.

Marinelli, Janet. 1980. "Eco-Crime on the Equator." *Environmental Action*, March.

McNeill, William H. 1976. *Plagues and Peoples*. New York: Anchor.

Myers, Norman. 1980. *Conversion of Tropical Moist Forests*. Washington, D.C.: National Academy of Sciences.

Has Success Spoiled the Crow?

Angell, Tony. 1978. *Ravens, Crows, Magpies, and Jays*. Seattle: University of Washington Press.

Coombs, Franklin. 1978. *The Crows: A Study of the Corvids of Europe*. London: B. T. Batsford.

Ficken, Millicent S. 1977. "Avian Play." *The Auk*, vol. 94, July.

Goodwin, Derek. 1976. *Crows of the World*. Ithaca: Comstock.

Simmons, K.E.L. 1957. "A Review of the Anting-Behaviour of Passerine Birds." *British Birds*, vol. 50, October.

———. 1966. "Anting and the Problem of Self-Stimulation." *Journal of Zoology*, vol. 149.

Wilmore, Sylvia Bruce. 1977. *Crows, Jays, Ravens and Their Relatives*. London: David and Charles.

The Widow Knows

Milzer, Albert. 1934. "On the Great Abundance of the Black Widow Spider." *Science*, vol. 80, November 2.

Thorop, Raymond W., and Weldon D. Woodson. 1976. *The Black Widow Spider*. New York: Dover.

The Troubled Gaze of the Octopus

Cousteau, Jacques-Yves, and Philippe Diolé. 1973. *Octopus and Squid: The Soft Intelligence*. Translated by J. F. Bernard. New York: Doubleday.

High, William L. 1976. "The Giant Pacific Octopus." *Marine Fisheries Review*, vol. 38, September.

Wells, M. J. 1978. *Octopus: Physiology and Behaviour of an Advanced Invertebrate*. London: Chapman and Hall.

Wells, Martin. 1983. "Cephalopods Do It Differently." *New Scientist*, vol. 100, November 3.

Avatars of the Soul in Malaya

Bänziger, Hans. 1968. "Preliminary Observations on a Skin-Piercing Blood-Sucking Moth (*Calyptra eustrigata* (Hymps.) (Lep., Noctuidae)) in Malaya." *Bulletin of Entomological Research*, vol. 58.

———. 1971. "Blood-sucking Moths of Malaya." *Fauna*, vol. 24.

Bänziger, H., and W. Buttiker. 1969. "Records of Eye-frequenting Lepidoptera from Man." *Journal of Medical Entomology*, vol. 6, January 30.

Borror, Donald J., and Dwight M. DeLong. 1971. *An Introduction to the Study of Insects*. New York: Holt, Rinehart and Winston.

Buttiker, W. 1959. "Blood-feeding Habits of Adult Noctuidae (Lepidoptera) in Cambodia." *Nature*, vol. 184, October 10.

Sandved, Kjell B. 1976. *Butterflies*. Photographs by Kjell B. Sandved, text by Jo Brewer. New York: Abrams.

Smart, Paul. 1976. *The Illustrated Encyclopedia of the Butterfly World*. London: Hamlyn.

Rumors of a Snake

Heuvelmans, Bernard. 1965. *On the Track of Unknown Animals*. Translated and abridged by Richard Garnett. New York: Hill and Wang.

Minton, Sherman A., Jr., and Madge Rutherford Minton. 1973. *Giant Reptiles*. New York: Scribner's.

Pope, Clifford H. 1962. *The Giant Snakes*. London: Routledge and Keegan Paul.

Wool of Bat

Allen, Glover Morrill. 1962. *Bats*. New York: Dover.

Barbour, Roger W., and Wayne H. Davis. 1969. *Bats of America*. Lexington: University Press of Kentucky.

Bat Research News (originally *Bat Banding News*), vols. 1–16, 1960–75. Compiled by Wayne H. Davis (Department of Zoology, University of Kentucky, Lexington) from founding to April 1970 and by Robert L. Martin (Department of Biology, University of Maine, Farmington) thereafter.

Feist, Joe Michael. 1982. "Bats Away!" *American Heritage*, April–May.

Peterson, Russell. 1964. *Silently, by Night*. New York: McGraw-Hill.

Wimsatt, William A., ed. 1970. *Biology of Bats*, vols. 1–2. New York: Academic.

Yalden, D. W., and P. A. Morris. 1975. *The Lives of Bats*. New York: Quadrangle.

The Excavation of Jack Horner

Bakker, Robert T. 1972. "Anatomical and Ecological Evidence of Endothermy in Dinosaurs." *Nature*, vol. 238, July 14.

———. 1975. "Dinosaur Renaissance." *Scientific American*, April.

Desmond, Adrian J. 1975. *The Hot-Blooded Dinosaurs*. New York: Dial.

Horner, John R. 1982. "Evidence of Colonial Nesting and 'Site Fidelity' Among Ornithischian Dinosaurs." *Nature*, vol. 297, June 24.

———. 1984. "The Nesting Behavior of Dinosaurs." *Scientific American*, April.

Horner, John R., and Robert Makela. 1979. "Nest of Juveniles Provides Evidence of Family Structure Among Dinosaurs." *Nature*, vol. 282, November 15.

Ostrom, John H. 1969. "Terrestrial Vertebrates as Indicators of Mesozoic Climates." In *Proceedings of the North American Paleontological Convention*.

The Lives of Eugène Marais

Ardrey, Robert. 1967. *African Genesis: A Personal Investigation into the Animal Origins and Nature of Man*. New York: Atheneum.

Marais, Eugène. 1973. *The Soul of the Ape*. With an introduction by Robert Ardrey. Harmondsworth: Penguin.

———. 1973. *The Soul of the White Ant*. With a biographical note by his son. Translated by Winifred de Kok. Harmondsworth: Penguin.

The Man with the Metal Nose

Dreyer, J.L.E. 1890. *Tycho Brahe*. Edinburgh: Adam and Charles Black.
Koestler, Arthur. 1963. *The Sleepwalkers: A History of Man's Changing Vision of the Universe*. New York: Grosset and Dunlap.
Murdin, Paul, and Leslie Murdin. 1978. *The New Astronomy*. New York: Crowell.
Ronan, Colin A. 1981. *The Practical Astronomer*. London: Roxby.

Animal Rights and Beyond

Lopez, Barry. 1983. "Renegotiating the Contracts." *Parabola*, spring.
Morris, Richard Knowles, and Michael W. Fox, eds. 1978. *On the Fifth Day: Animal Rights and Human Ethics*. Washington, D.C.: Acropolis.
Regan, Tom. 1982. *All That Dwells Therein: Animal Rights and Environmental Ethics*. Berkeley: University of California Press.
———. 1983. *The Case for Animal Rights*. Berkeley: University of California Press.
Regan, Tom, and Peter Singer, eds. 1976. *Animal Rights and Human Obligations*. Englewood Cliffs, N.J.: Prentice Hall.
Singer, Peter. 1975. *Animal Liberation: A New Ethics for Our Treatment of Animals*. New York: New York Review Books.

The Tree People

Beasley, R. S., and J. O. Klemmedson. 1973. "Recognizing Site Adversity and Drought-Sensitive Trees in Stands of Bristlecone Pine (*Pinus longaeva*)." *Economic Botany*, vol. 27, January-March.
Currey, Donald R. 1965. "An Ancient Bristlecone Pine Stand in Eastern Nevada." *Ecology*, vol. 46, early summer.
Ferguson, C. W. 1968. "Bristlecone Pine: Science and Aesthetics." *Science*, vol. 159, February 23.
Fowles, John. 1979. *The Tree*. Photographs by Frank Horvat. Boston: Little, Brown.
LaMarche, Valmore C., Jr. 1969. "Environment in Relation to Age of Bristlecone Pines." *Ecology*, vol. 50, winter 1969.
Rogers, Julia Ellen. 1922. *The Tree Book*. New York: Doubleday, Page.

Love's Martyrs

Childerhose, R. J., and Marj Trim. 1979. *Pacific Salmon and Steelhead Trout*. Seattle: University of Washington Press.
Gadgil, Mahdav, and William H. Bossert. 1970. "Life Historical Consequences of Natural Selection." *American Naturalist*, vol. 104, January-February.
Gentry, Howard Scott. 1982. *Agaves of Continent North America*. Tucson: University of Arizona Press.
Janzen, Daniel H. 1976. "Why Bamboos Wait So Long to Flower." *Annual Review of Ecology and Systematics*, vol. 7.
Netboy, Anthony. 1974. *The Salmon: Their Fight for Survival*. Boston: Houghton Mifflin.

Schaffer, William M. 1974. "Selection for Optimal Life Histories: The Effects of Age Structure." *Ecology*, vol. 55, early spring.

Schaffer, William M., and Michael L. Rosenzweig. 1977. "Selection for Optimal Life Histories. II: Multiple Equilibria and the Evolution of Alternative Reproductive Strategies." *Ecology*, vol. 58, winter.

Schaffer, William M., and M. Valentine Schaffer. 1979. "The Adaptive Significance of Variations in Reproductive Habit in the Agavaceae. II: Pollinator Foraging Behavior and Selection for Increased Reproductive Expenditure." *Ecology*, vol. 60, October.

A Deathly Chill

I benefited from—in addition to the Ted Lathrop pamphlet and the Associated Press story—the help of Mark Smith, a reporter for the *Tri-County Tribune* of Deer Park, Washington, not far from Chattaroy. Mr. Smith shared with me his coverage of the Ram Patrol misfortune.

Is Sex Necessary?

Beatty, R. A. 1957. *Parthenogenesis and Polyploidy in Mammalian Development*. Cambridge: Cambridge University Press.

Birky, C. William, Jr., and John J. Gilbert. 1971. "Parthenogenesis in Rotifers: The Control of Sexual and Asexual Reproduction." *American Zoologist*, vol. 11.

Blackmun, Roger. 1974. *Aphids*. London: Ginn.

Ghiselin, Michael T. 1974. *The Economy of Nature and the Evolution of Sex*. Berkeley: University of California Press.

Lowe, A. D., ed. 1973. *Perspectives in Aphid Biology*. Auckland: The Entomological Society of New Zealand.

Suomalainen, Esko. 1961. "Significance of Parthenogenesis in the Evolution of Insects." *Annual Review of Entomology*, vol. 7.

White, M.J.D. 1954. *Animal Cytology and Evolution*. Cambridge: Cambridge University Press.

Desert Sanitaire

Abbey, Edward. 1977. *The Journey Home: Some Words in Defense of the American West*. New York: Dutton.

George, Uwe. 1977. *In the Deserts of This Earth*. Translated by Richard and Clara Winston. New York: Harcourt Brace Jovanovich.

Jaeger, Edmund C. 1957. *The North American Deserts*. Stanford, Calif.: Stanford University Press.

Krutch, Joseph Wood. 1980. *The Voice of the Desert: A Naturalist's Interpretation*. New York: Morrow.

Petrov, M. P. 1966. *The Deserts of Central Asia*. Washington, D.C.: Joint Publications Research Service.

Pickwell, Gayle. 1939. *Deserts*. New York: McGraw-Hill.

Jeremy Bentham, the *Pietà,* and a Precious Few Grayling

Holton, George D. 1971. "Montana Grayling: The Lady of the Streams." *Montana Outdoors*, September-October. For facts and numbers concerning the hatchery and planting program in Montana, I am also indebted to George Holton (personal communication), to a fact sheet prepared by Bill Gould, and to a typescript report on Montana grayling habitat done by Earl E. Willard and Margaret Herman for the U.S. Forest Service.

Kruse, Thomas E. 1958. "Grayling of Grebe Lake, Yellowstone National Park, Wyo." *Fishery Bulletin of the Fish and Wildlife Service*, vol. 59.

Nelson, Perry H. 1954. "Life History and Management of the American Grayling (*Thymallus signifier tricolor*) in Montana." *Journal of Wildlife Management*, vol. 18, July.

Yin and Yang in the Tularosa Basin

Allmendinger, Roger J. n.d. "Hydrologic Control over the Origin of Gypsum at Lake Lucero, White Sands National Monument, New Mexico." Unpublished master's thesis in the library at the headquarters of White Sands National Monument.

Emerson, Fred W. 1935. "An Ecological Reconaissance in the White Sands, New Mexico." *Ecology*, vol. 16, April.

Goodchild, Peter. 1981. *J. Robert Oppenheimer: Shatterer of Worlds*. Boston: Houghton Mifflin.

Groueff, Stéphane. 1967. *Manhattan Project: The Untold Story of the Making of the Atomic Bomb*. Boston: Little, Brown.

Groves, Leslie R. 1975. *Now It Can Be Told: The Story of the Manhattan Project*. New York: Da Capo.

Harrington, M. W. 1885. "Lost Rivers." *Science*, vol. 6, September 25.

Reid, William H., project director. 1980. *Final Report: White Sands National Monument Natural Resources and Ecosystem Analysis*. El Paso: Research Report Number 12, Laboratory for Environmental Biology, University of Texas at El Paso.

Schaafsma, Polly. 1975. *Rock Art in New Mexico*. Albuquerque: University of New Mexico Press.

Smith, Alice Kimball, and Charles Weiner, eds. 1980. *Robert Oppenheimer: Letters and Recollections*. Cambridge, Mass.: Harvard University Press.

Sutherland, Kay. n.d. "Petroglyphs at Three Rivers, New Mexico: A Partial Survey." *The Artifact* (published by the El Paso Archeological Society), vol. 16.

Planet of Weeds

Aldous, Peter. 1993. "Tropical Deforestation: Not Just a Problem in Amazonia." *Science*, vol. 259, March 5

Alverson, William S., Donald M. Waller, and Stephen L. Solheim. 1988. "Forests Too Deer: Edge Effects in Northern Wisconsin." *Conservation Biology*, vol. 2, no. 4.

Anonymous (signed C. H.). 1974. "Scientists Talk of the Need for Conservation and an Ethic of Biotic Diversity to Slow Species Extinction." *Science*, vol. 184, May 10.

Anonymous. 1990. *The Economist Book of Vital World Statistics: A Complete Guide to the World in Figures*. London: Hutchinson.

Athanasiou, Tom. 1996. *Divided Planet: The Ecology of Rich and Poor.* Boston: Little, Brown.

Babbitt, Bruce. 1998. "Kudzu, Kudzu, Kill! Kill! Kill!" Excerpt from a speech to the Science in Wildland Weed Management Symposium, April 8. Reprinted (not his choice of title, presumably) in *Harper's*, July.

Barber, Benjamin R. 1996. *Jihad vs. MacWorld.* New York: Ballantine.

Baskin, Yvonne. 1996. "Curbing Undesirable Invaders." *BioScience*, vol. 46, no. 10.

————. 1997. *The Work of Nature: How the Diversity of Life Sustains Us.* Washington, D.C.: Island.

Botkin, Daniel B. 1990. *Discordant Harmonies: A New Ecology for the Twenty-first Century.* New York: Oxford University Press.

Calvin, William H. 1998. "The Great Climate Flip-flop." *Atlantic Monthly*, January.

Challoner, W. G., and A. Hallam, eds. 1989. *Evolution and Extinction.* Cambridge: Cambridge University Press.

Cohen, Joel E. 1995. *How Many People Can the Earth Support?* New York: W. W. Norton.

————. 1995. "Population Growth and Earth's Human Carrying Capacity." *Science*, vol. 269, July 21.

Committee for the Compilation of Materials on Damage Caused by the Atomic Bombs in Hiroshima and Nagasaki. 1981. *Hiroshima and Nagasaki: The Physical, Medical, and Social Effects of the Atomic Bombings.* Translated by Eisei Ishikawa and David L. Swain. New York: Basic.

Daily, Gretchen, ed. 1997. *Nature's Secrets: Societal Dependence on Natural Ecosystems.* Washington, D.C.: Island.

Dobson, Andrew P. 1996. *Conservation and Biodiversity.* New York: Scientific American Library.

Donovan, Stephen K., ed. 1989. *Mass Extinctions: Processes and Evidence.* New York: Columbia University Press.

Durning, Alan Thein. 1994. "The Conundrum of Consumption." In Mazur (1994) and adapted from Durning (1992), *How Much Is Enough? The Consumer Society and the Fate of the Earth*, New York: W. W. Norton.

Easterbrook, Gregg. 1995. *A Moment on the Earth: The Coming Age of Environmental Optimism.* New York: Viking.

Ehrenfeld, David W. 1970. *Biological Conservation.* New York: Holt, Rinehart and Winston.

Ehrlich, Paul, and Anne Ehrlich. 1981. *Extinction: The Causes and Consequences of the Disappearance of Species.* New York: Random House.

Ehrlich, Paul R. 1986. "Extinction: What Is Happening Now and What Needs to be Done." In Elliott (1986).

Eldredge, Niles. 1998. *Life in the Balance: Humanity and the Biodiversity Crisis.* Princeton, N.J.: Princeton University Press.

Elliott, David K., ed. 1986. *Dynamics of Extinction.* New York: Wiley.

Elton, Charles. (1958) 1977. *The Ecology of Invasions by Animals and Plants.* London: Chapman and Hall.

Engelman, Robert, and Richard P. Cincotta. 1997. "Nature Displaced: Human Population Trends, Projections, and Their Meanings." Draft of a paper presented to Nature and Human Society: The Quest for a Sustainable World, a conference held at the National Academy of Sciences, Washington, D.C., October.

Erwin, Douglas H. 1994. "The Permo-Triassic Extinction." *Nature*, vol. 367, January 20,.

Feduccia, Alan. 1996. *The Origin and Evolution of Birds.* New Haven, Conn.: Yale University Press.

Flack, Stephanie, and Elaine Furlow. 1996. "America's Least Wanted." *Nature Conservancy*, November-December.

Fukuyama, Francis. 1992. *The End of History and the Last Man.* New York: Avon.

Goldschmidt, Tijs. 1996. *Darwin's Dreampond: Drama in Lake Victoria.* Translated by Sherry Marx-Macdonald. Cambridge, Mass.: MIT Press.

Goodman, Steven M., and Bruce D. Patterson, eds. 1997. *Natural Change and Human Impact on Madagascar.* Washington, D.C.: Smithsonian Institution Press.

Greider, William. 1998. *One World, Ready or Not: The Manic Logic of Global Capitalism.* New York: Touchstone.

Harrison, Paul. 1992. *The Third Revolution: Environment, Population and a Sustainable World.* London: I. B. Tauris.

Heaney, Lawrence R., and Jacinto C. Regalado, Jr. 1998. *Vanishing Treasures of the Philippine Rain Forest.* Chicago: Field Museum.

Herman, Arthur. 1997. *The Idea of Decline in Western History.* New York: Free Press.

Homer-Dixon, Thomas F. 1993. *Environmental Scarcity and Global Security.* Headline Series #300. New York: Foreign Policy Association.

Horgan, John. 1997. *The End of Science: Facing the Limits of Knowledge in the Twilight of the Scientific Age.* New York: Broadway Books.

Jablonski, David. 1986. "Background and Mass Extinctions: The Alternation of Macroevolutionary Regimes." *Science*, vol. 231, January 10.

———. 1986. "Causes and Consequences of Mass Extinctions: A Comparative Approach." In Elliott (1986).

———. 1986. "Mass Extinctions: New Answers, New Questions." In Kaufman and Mallory (1986).

———. 1989. "The Biology of Mass Extinction: A Paleontological View." In Challoner and Hallam (1989).

———. 1991. "Extinctions: A Paleontological Perspective." *Science*, vol. 253, August 16.

———. 1996. "Mass Exntinctions: Persistent Problems and New Directions." In Ryder, G., D. Fastovsky, and S. Gartner, eds. *The Cretaceous-Tertiary Event and Other Catastrophes in Earth History.* Special Paper 307. Boulder, Colo.: Geological Society of America.

Jablonski, David, Douglas H. Ervin, and Jere H. Lipps, eds. 1996. *Evolutionary Paleobiology.* Chicago: University of Chicago Press.

Jablonski, David, and David M. Raup. 1995. "Selectivity of End-Cretaceous Marine Bivalve Extinctions." *Science*, vol. 268, April 21.

Kaplan, Robert D. 1994. "The Coming Anarchy." *Atlantic Monthly*, February.

Kaufman, Les, and Kenneth Mallory, eds. 1986. *The Last Extinction*. Cambridge, Mass.: MIT Press.

Keck, Andrew, Narendra P. Sharma, and Gershon Feder. 1994. "Population Growth, Shifting Cultivation, and Unsustainable Agricultural Development: A Case Study in Madagascar." World Bank Discussion Papers No. 234. Washington, D.C.: World Bank.

Kramer, Randall, Carel van Schaik, and Julie Johnson. 1997. *Last Stand: Protected Areas and the Defense of Tropical Biodiversity*. New York: Oxford University Press.

Lapham, Lewis H., ed. 1997. *History: The End of the World*. New York: History Book Club.

Lawton, John H., and Robert M. May, eds. 1995. *Extinction Rates*. New York: Oxford University Press.

Leakey, Richard, and Roger Lewin. 1995. *The Sixth Extinction: Patterns of Life and the Future of Humankind*. New York: Doubleday.

Leslie, John. 1996. *The End of the World: The Science and Ethics of Human Extinction*. London: Routledge.

Lovejoy, Thomas E. 1980. "A Projection of Species Extinctions." In *The Global 2000 Report to the President*. Washington, D.C.: Council on Environmental Quality.

———. 1997. "National Security, National Interest and Sustainability." Draft of a paper presented to the Nature and Human Society symposium, Washington, D.C., October.

Mann, Charles C. 1991. "Extinction: Are Ecologists Crying Wolf?" *Science*, vol. 253, August 16,.

Martin, Paul S., and Richard G. Klein. 1984. *Quaternary Extinctions: A Prehistoric Revolution*. Tucson: University of Arizona Press.

Mazur, Laurie Ann, ed. 1994. *Beyond the Numbers: A Reader on Population, Consumption, and the Environment*. Washington, D.C.: Island.

McDonald, Kim A. 1997. "Scientists Refine Estimates of Number of Species and Their Rate of Extinction." *Chronicle of Higher Education*, November 14.

McKibben, Bill. 1989. *The End of Nature*. New York: Random House.

———. 1998. "A Special Moment in History." *Atlantic Monthly*, May.

McKnight, Bill N. 1993. *Biological Pollution: The Control and Impact of Invasive Exotic Species*. Indianapolis: Indiana Academy of Science.

McNeill, William H. 1992. *The Global Condition: Conquerors, Catastrophes, and Community*. Princeton, N.J.: Princeton University Press.

Myers, Norman. 1976. "An Expanded Approach to the Problem of Disappearing Species." *Science*, vol. 193, July 16.

———. 1979. *The Sinking Ark: A New Look at the Problem of Disappearing Species*. New York: Pergamon.

———. 1980. *Conversion of Tropical Moist Forests*. Washington, D.C.: National Academy of Sciences.

———. 1996. *Ultimate Security: The Environmental Basis of Political Stability*. Washington, D.C.: Island.

Nitecki, Matthew H., ed. 1984. *Extinctions*. Chicago: University of Chicago Press.

Office of Technology Assessment, U. S. Congress. 1993. *Harmful Non-Indigenous Species in the United States*. Washington, D.C.: U.S. Government Printing Office.

Peters, Robert L., and Thomas E. Lovejoy, eds. 1992. *Global Warming and Biological Diversity*. New Haven, Conn.: Yale University Press.

Pimm, Stuart, Gareth Russell, John L. Gittleman, and Thomas M. Brooks. 1995. "The Future of Biodiversity." *Science*, vol. 269, July 21.

Pimm, Stuart L. 1991. *The Balance of Nature?: Ecological Issues on the Conservation of Species and Communities*. Chicago: University of Chicago Press.

Pimm Stuart L., and Thomas M. Brooks. 1997. "The Sixth Extinction: How Large, How Soon, and Where?" Draft of a paper presented to the Nature and Human Society symposium, Washington, D.C. October.

Pimm, Stuart L., and John H. Lawton. 1998. "Planning for Biodiversity." *Science*, vol. 279, March 27.

Potts, Rick. 1996. *Humanity's Descent: The Consequences of Ecological Instability*. New York: Morrow.

Raup, David M. 1986. *The Nemesis Affair: A Story of the Death of Dinosaurs and the Ways of Science*. New York: W. W. Norton.

———. 1991. *Extinction: Bad Genes or Bad Luck?* New York: W. W. Norton.

Sepkoski, J. John, Jr., and David M. Raup. 1986. "Periodicity in Marine Extinction Events." In Elliott (1986).

Simberloff, Daniel. 1986. "Are We on the Verge of a Mass Extinction in Tropical Rain Forests?" In Elliott (1986).

Simon, Julian L. 1981. *The Ultimate Resource*. Princeton, N.J.: Princeton University Press.

———. 1986. "Disappearing Species, Deforestation and Data." *New Scientist*, May 19.

Simon, Julian L., and Aaron Wildavsky. 1993. "Facts, Not Species, Are Imperiled." *New York Times*, May 13.

Soulé, Michael E., and M. A. Sanjayan. 1998. "Conservation Targets: Do They Help?" *Science*, vol. 279, March 27.

Stanley, Steven M. 1987. *Extinction*. New York: Scientific American Library.

Steadman, David W. 1995. "Prehistoric Extinctions of Pacific Island Birds: Biodiversity Meets Zooarchaeology." *Science*, vol. 267, February 24.

Terborgh, John, and Carel van Schaik. 1997. "Minimizing Species Loss: The Imperative of Protection." In Kramer et al. (1997).

Tudge, Colin. 1996. *The Time Before History: 5 Million Years of Human Impact*. New York: Scribner.

United Nations Secretariat, Population Division. 1998. *World Population Projections to 2150*. New York: United Nations.

Ward, Peter. 1994. *The End of Evolution: A Journey in Search of Clues to the Third Mass Extinction Facing Planet Earth*. New York: Bantam.

Wells, H. G. (1895) 1992. *The Time Machine*. New York: Tom Doherty Associates.

Western, David, and Mary C. Pearl, eds. 1989. *Conservation for the Twenty-first Century*. New York: Oxford University Press.

Whitmore, T. C., and J. A. Sayer. 1992. *Tropical Deforestation and Species Extinction*. London: Chapman and Hall.

Wilson, Edward O. 1992. *The Diversity of Life*. Cambridge, Mass.: Belknap/Harvard University Press.

World Conservation Monitoring Center. 1990. *1990 IUCN Red List of Threatened Animals*. Gland, Switzerland: International Union for the Conservation of Nature and Natural Resources.

The River Jumps Over the Mountain

Auden, W. H. 1989. *Selected Poems*. Edited by Edward Mendelson. New York: Vintage.

Beus, Stanley S., and Michael Morales, eds. 1990. *Grand Canyon Geology*. New York: Oxford University Press.

Dean, Dennis R. 1992. *James Hutton and the History of Geology*. Ithaca, N.Y.: Cornell University Press.

Gould, Stephen Jay. 1987. *Time's Arrow, Time's Cycle: Myth and Metaphor in the Discovery of Geological Time*. Cambridge, Mass.: Harvard University Press.

Hamblin, W. Kenneth, and J. Keith Rigby. 1969. *Guidebook to the Colorado River, Part 2: Phantom Ranch in Grand Canyon National Park to Lake Mead, Arizona-Nevada*. Brigham Young University Geology Studies, vol. 16. Provo, Utah: Department of Geology, Brigham Young University.

McPhee, John. 1981. *Basin and Range*. New York: Farrar, Straus and Giroux.

Playfair, John. (1802) 1964. *Illustrations of the Huttonian Theory of the Earth*. Facsimile reprint, with an introduction by George W. White. New York: Dover.

Price, L. Greer. 1999. *An Introduction to Grand Canyon Geology*. Grand Canyon, Ariz.: Grand Canyon Association.

Stevens, Larry. 1999. *The Colorado River in Grand Canyon: A Comprehensive Guide to Its Natural and Human History*. Flagstaff, Ariz.: Red Lake.

Whitney, Stephen R. 1996. *A Field Guide to the Grand Canyon*. Seattle: Mountaineers.

The Post-Communist Wolf

Behr, Edward. 1991. *Kiss the Hand You Cannot Bite: The Rise and Fall of the Ceauşescus*. New York: Villard.

Codrescu, Andrei. 1991. *The Hole in the Flag: A Romanian Exile's Story of Return and Revolution*. New York: William Morrow.

Crişan, Vasile. 1994. *Jäger? Schlächter: Ceauşescu*. (Privately translated for me by Eduard Érsek as "Ceauşescu: Hunter or Butcher?") Mainz: Verlag Dieter Hoffmann.

Cullen, Robert. 1990. "Report from Romania." *The New Yorker*, April 2.

Deletant, Dennis. 1995. *Ceauşescu and the Securitate: Coercion and Dissent in Romania, 1965–1989*. London: Hurst.

Fischer-Galaţi, Stephen. 1970. *Twentieth-Century Rumania*. New York: Columbia University Press.

Georgescu, Vlad. 1991. *The Romanians: A History*. Translated by Alexandra Bley-Vroman, edited by Matei Calinescu. Columbus: Ohio State University Press.

Hale, Julian. 1971. *Ceauşescu's Romania: A Political Documentary*. London: George G. Harrap.

Judt, Tony. "Romania: Bottom of the Heap." *New York Review of Books*, November 1.

Mech, L. David. 1981. *The Wolf: The Ecology and Behavior of an Endangered Species*. Minneapolis: University of Minnesota Press.

Mertens, Annette, and Christoph Promberger. 2000. "Economic Aspects of Large Carnivore-Livestock Conflicts in Romania." (Draft.)

Pacepa, Lieutenant General Ion Mihai. 1987. *Red Horizons: The True Story of Nicolae and Elena Ceauşescus'* [sic] *Crimes, Lifestyle, and Corruption*. Washington, D.C.: Regnery Gateway.

The Megatransect

Ambrose, Stephen E. 1996. *Undaunted Courage: Meriwether Lewis, Thomas Jefferson, and the Opening of the American West*. New York: Touchstone.

Barnes, R.F.W., and S. A. Lahm. 1997. "An Ecological Perspective on Human Densities in the Central African Forests." *Journal of Applied Ecology*, vol. 34.

Farrell, Byron. 1985. *The Man Who Presumed: A Biography of Henry M. Stanley*. New York: W. W. Norton.

Fay, J. Michael. 1997. "The Ecology, Social Organization, Populations, Habitat and History of the Western Lowland Gorilla (*Gorilla gorilla gorilla* Savage and Wyman 1847)." Unpublished doctoral dissertation, Washington University, St. Louis.

Georges, Alain-Jean, Eric M. Leroy, André A. Renaut, Carol Tevi Benissan, René J. Nabias, Minh Trinh Ngoc, Paul I. Obiang, J.P.M. Lepage, Eric J. Bertherat, David D. Bénoni, E. Jean Wickings, Jacques P. Amblard, Joseph M. Lansoud-Soukate, J. M. Milleliri, Sylvain Baize, and Marie-Claude Georges-Courbot. 1999. "Ebola Hemorrhagic Fever Outbreaks in Gabon, 1994–1997: Epidemiologic and Health Control Issues." *Journal of Infectious Diseases*, vol. 1, no. 79, supplement 1.

Huijbregts, Bas. 2000. "Gorilles et Chimpanzees a Minkebe: Decimes par Ebola?" Unpublished report to the World Wildlife Fund, February 9.

Kingdon, Jonathan. 1997. *The Kingdon Field Guide to African Mammals*. New York: Academic.

Lahm, Sally. 1993. "Ecology and Economics of Human/Wildlife Interaction in Northeastern Gabon." Unpublished doctoral dissertation, New York University, New York.

McLynn, Frank. 1992. *Hearts of Darkness: The European Exploration of Africa*. New York: Carroll and Graf.

Oslisly, Richard. 1994. "The Middle Ogooué Valley: Cultural Changes and Paleoclimatic Implications of the Last Four Millennia." *Azania*, vols. 29–30: A special volume on "The Growth of Farming Communities in Africa from the Equator Southwards," J.E.G. Sutton, ed. The British Institute in Eastern Africa.

Tutin, C.E.G., and M. Fernandez. 1984. "Nationwide Census of Gorilla (*Gorilla g. gorilla*) and Chimpanzee (*Pan t. troglodytes*) Populations in Gabon." *American Journal of Primatology*, vol. 6.

Vansina, Jan. 1990. *Paths in the Rainforests: Toward a History of Political Tradition in Equatorial Africa*. Madison: University of Wisconsin Press.

West, Richard. 1972. *Brazza of the Congo: European Exploration and Exploitation in French Equatorial Africa*. London: Jonathan Cape.

A Passion for Order

The librarians, archivists, and other officials at the Linnean Society of London—notably Gina Douglas, Lynda Brooks, and the society's executive secretary, Adrian Thomas—were extremely hospitable to my research for this piece, offering me access to Linnaeus's personal papers and collections. In Uppsala, Mats Block and Mikael Norrby, as well as Karin Martinsson, Carl-Olof Jacobson, and many other people, welcomed my visits to Linnaeus's houses and my persistent questions. Peter Raven, in e-mail exchanges, also helped guide my understanding of Linnaeus's contribution to biology.

Blackwelder, R. E., and Alan Boyden. 1952. "The Nature of Systematics." *Systematic Zoology*, vol. 1, no. 1, spring 1952.

Blunt, Wilfrid. 2001. *Linnaeus: The Compleat Naturalist*. Princeton, N.J.: Princeton University Press.

Frängsmyr, Tore, ed. 1994. *Linnaeus: The Man and His Work*. Canton, Mass.: Watson.

Linnaeus, Carl. (1751) 2003. *Linnaeus' Philosophia Botanica*. Translated by Stephen Freer. New York: Oxford University Press.

Mayr, Ernst. 1982. *The Growth of Biological Thought: Diversity, Evolution, and Inheritance*. Cambridge, Mass.: Belknap/Harvard University Press.

Müller-Wille, Steffan. 2006. "Linnaeus' Herbarium Cabinet: A Piece of Furniture and Its Function." *Endeavour*, vol. 30, June.

Raven, Peter H., Brent Berlin, and Dennis E. Breedlove. 1971. "The Origins of Taxonomy." *Science*, vol. 174, December 17.

Raven, Peter H., Ray F. Evert, and Susan E. Eichhorn. 1992. *Biology of Plants*. New York: Worth.

Reeds, Karen. 2004. "When the Botanist Can't Draw: The Case of Linnaeus." *Interdisciplinary Science Reviews*, vol. 29, no. 3.

Ross, Herbert H. 1974. *Biological Systematics*. Reading, Mass.: Addison-Wesley.

Stearn, W. T. 1959. "The Background of Linnaeus's Contributions to the Nomenclature and Methods of Systematic Biology." *Systematic Zoology*, vol. 8, no. 1, March.

Citizen Wiley

Thomas, Dylan. 1933. "And Death Shall Have No Dominion." *New English Weekly*, March. Reprinted in his *Twenty-Five Poems* (1936) and in my copy of *Modern American Poetry/Modern British Poetry* (1958), given to me in 1966 by a friend who is also now dead but not forgotten.

Clone Your Troubles Away

Alexander, Brian. 2004. "John Sperling Wants You to Live Forever." *Wired,* February.

Bawa, Kamaljit S., Shaily Menon, and Leah R. Gorman. 1997. "Cloning and Conservation of Biological Diversity: Paradox, Panacea, or Pandora's Box?" *Conservation Biology*, vol. 11, no. 4, August.

Commoner, Barry. 2002. "Unraveling the DNA Myth: The Spurious Foundation of Genetic Engineering." *Harper's*, February.

Corley-Smith, Graham E., and Bruce P. Brandhorst. 1999. "Preservation of Endangered Species and Populations: A Role for Genome Banking, Somatic Cell Cloning, and Androgenesis?" *Molecular Reproduction and Development*, vol. 53.

Cohen, Jon. 1997. "Can Cloning Help Save Beleaguered Species?" *Science*, vol. 276, May 30.

Gomez, Martha, Earle Pope, Rebecca Harris, Susan Mikota, and Betsy L. Dresser. 2003. "Development of In Vitro Matured, In Vitro Fertilized Domestic Cat Embryos Following Cryopreservation, Culture and Transfer." *Theriogenology*, vol. 60, issue 2, July.

Gomez, Martha C., Jill A. Jenkins, Angelica Giraldo, Rebecca F. Harris, Amy King, Betsy L. Dresser, and Charles Earle Pope. 2003. "Nuclear Transfer of Synchronized African Wild Cat Somatic Cells into Enucleated Domestic Cat Oocytes." *Biology of Reproduction*, vol. 69.

Graeber, Charles. 2000. "How Much Is That Doggy In Vitro?" *Wired*, March.

Kraemer, Duane C., Gary T. Moore, and Martin A. Kramen. 1976. "Baboon Infant Produced by Embryo Transfer." *Science*, vol. 192, June 18.

Lanza, Robert P., Jose B. Cibelli, Francisca Diaz, Carlos T. Moraes, Peter W. Farin, Charlotte E. Farin, Carolyn J. Hammer, Michael D. West, and Philip Damiani. 2000. "Cloning of an Endangered Species (*Bos gaurus*) Using Interspecies Nuclear Transfer." *Cloning*, vol. 2, no. 2.

Lanza, Robert P., Jose B. Cibelli, David Faber, Raymond W. Sweeney, Boyd Henderson, Wendy Nevala, Michael D. West, and Peter J. Wettstein. 2001. "Cloned Cattle Can Be Healthy and Normal." *Science*, vol. 294, November 30.

Lanza, Robert P., Betsy L. Dresser, and Philip Damiani. 2000. "Cloning Noah's Ark." *Scientific American*, November.

Loi, Pasqualino, Grazyna Ptak, Barbara Barboni, Josef Fulka, Jr., Pietro Cappai, and Michael Clinton. 2001. "Genetic Rescue of an Endangered Mammal by Cross-Species Nuclear Transfer Using Post-Mortem Somatic Cells." *Nature Biotechnology*, vol. 19, no. 10, October.

Long, C. R., S. C. Walker, R. T. Tang, and M. E. Westhusin. 2003. "New Commercial Opportunities for Advanced Reproductive Technologies in Horses, Wildlife, and Companion Animals." *Theriogenology*, vol. 59.

McGrath, James, and Davor Solter. 1984. "Inability of Mouse Blastomere Nuclei Transferred into Enucleated Zygotes to Support Development In Vitro." *Science*, vol. 226, December 14.

Rennie, John. 2000. "Cloning and Conservation." *Scientific American*, November.

Rieseberg, Loren H., Barry Sinervo, C. Randal Linder, Mark C. Ungerer, and Dulce M. Arias. 1996. "Role of Gene Interactions in Hybrid Speciation: Evidence from Ancient and Experimental Hybrids." *Science*, vol. 272, May 3.

Sandel, Michael J. 2004. "The Case Against Perfection." *Atlantic Monthly*, April.

Shin, Taeyoung, Duane Kraemer, Jane Pryor, Ling Liu, James Rugila, Lisa Howe, Sandra Buck, Keith Murphy, Leslie Lyons, and Mark Westhusin. 2002. "A Cat Cloned by Nuclear Transplantation." *Nature*, vol. 415, February 21.

Stone, Richard. 1999. "Cloning the Woolly Mammoth." *Discover*, April.

Weidensaul, Scott. 2002. "Raising the Dead." *Audubon*, May-June.

Westhusin, Mark, Katrin Hinrichs, Young-ho Choi, Taeyoung Shin, Ling Liu, and Duane Kraemer. 2003. ""Cloning Companion Animals (Horses, Cats, and Dogs)." *Cloning and Stem Cells*, vol. 5, no. 4.

Westhusin, Mark, and Jorge Piedrahita. 2000. "Three Little Pigs Worth the Huff and Puff?" *Nature Biotechnology*, vol. 18, November.

Westhusin, M. E., R. C. Burghardt, J. N. Rugila, L. A. Willingham, L. Liu, T. Shin, L. M. Howe, D. C. Kraemer. 2001. "Potential for Cloning Dogs." *Journal of Reproduction and Fertility, Supplement*, vol. 59.

Westhusin, M. E., C. R. Long, T. Shin, J. R. Hill, C. R. Looney, J. H. Pryor, and J. A. Piedrahita.. 2001. *Theriogenology*, vol. 55.

Williams, B., T. Shin, L. Liu, G. Flores-Foxworth, J. Romano, M. Westhusin, and D. Kraemer. 2002. "Interspecies Nuclear Transfer of Desert Bighorn Sheep (*Ovis canadensis mexicana*)." *Theriogenology*, vol. 57, January 1.

INDEX

David Quammen is the author of four books of fiction and seven acclaimed nonfiction titles, including *The Reluctant Mr. Darwin* and *The Song of the Dodo,* which was awarded the John Burroughs Medal for natural history writing. He has been honored with an Academy Award in Literature from the American Academy of Arts and Letters and is a three-time recipient of the National Magazine Award, most recently for a cover story in *National Geographic* entitled "Was Darwin Wrong?" Quammen currently holds the Wallace Stegner Chair of Western American Studies at Montana State University, in Bozeman. He is also a contributing writer for *National Geographic.*